Springer Theses

Recognizing Outstanding Ph.D. Research

Aims and Scope

The series "Springer Theses" brings together a selection of the very best Ph.D. theses from around the world and across the physical sciences. Nominated and endorsed by two recognized specialists, each published volume has been selected for its scientific excellence and the high impact of its contents for the pertinent field of research. For greater accessibility to non-specialists, the published versions include an extended introduction, as well as a foreword by the student's supervisor explaining the special relevance of the work for the field. As a whole, the series will provide a valuable resource both for newcomers to the research fields described, and for other scientists seeking detailed background information on special questions. Finally, it provides an accredited documentation of the valuable contributions made by today's younger generation of scientists.

Theses may be nominated for publication in this series by heads of department at internationally leading universities or institutes and should fulfill all of the following criteria

- They must be written in good English.
- The topic should fall within the confines of Chemistry, Physics, Earth Sciences, Engineering and related interdisciplinary fields such as Materials, Nanoscience, Chemical Engineering, Complex Systems and Biophysics.
- The work reported in the thesis must represent a significant scientific advance.
- If the thesis includes previously published material, permission to reproduce this must be gained from the respective copyright holder (a maximum 30% of the thesis should be a verbatim reproduction from the author's previous publications).
- They must have been examined and passed during the 12 months prior to nomination.
- Each thesis should include a foreword by the supervisor outlining the significance of its content.
- The theses should have a clearly defined structure including an introduction accessible to new PhD students and scientists not expert in the relevant field.

Indexed by zbMATH.

Giulia Di Gregorio

A Beauty-ful Boson

Search for the Higgs Boson Produced in Association with a Vector Boson and Decaying into a Pair of *b*-quarks Using Large-*R* Jets with the ATLAS Detector

Doctoral Thesis accepted by
University of Pisa, Pisa, Italy

Springer

Author
Giulia Di Gregorio
University of Pisa
Pisa, Italy

Supervisors
Dr. Fabrizio Scuri
Istituto Nazionale di Fisica Nucleare
(INFN), Sezione di Pisa
Pisa, Italy

Dr. Paolo Francavilla
Istituto Nazionale di Fisica Nucleare
(INFN), Sezione di Pisa
Pisa, Italy

ISSN 2190-5053 ISSN 2190-5061 (electronic)
Springer Theses
ISBN 978-3-031-20012-0 ISBN 978-3-031-20013-7 (eBook)
https://doi.org/10.1007/978-3-031-20013-7

This Springer imprint is published by the registered company Springer Nature Switzerland AG
The registered company address is: Gewerbestrasse 11, 6330 Cham, Switzerland

To Denis,
for being a constant source of support and encouragement

Supervisors' Foreword

Giulia Di Gregorio's thesis is part of a large effort of the ATLAS experiment at CERN to extensively study the decay of the Higgs boson into a pair of b-quarks. Due to the large branching fraction of the Higgs boson decay into b-quarks, large samples of events can be selected and used to study both the Higgs production mechanisms and its coupling to heavy quarks. Moreover, specific studies on the distribution shape of some kinematical observables of the events can lead to the discovery of possible contributions due to new phenomena from physics processes other than those foreseen in the Standard Model, the current theoretical framework describing the main interactions among elementary particles.

This thesis focus is on the study of the events in which a high momentum Higgs boson is produced in proton-proton collisions in association with a W boson or with a Z boson. The W/Z bosons are selected by reconstructing their lepton decay modes with event categorization based on the number of charged leptons in the final states.

At high momentum, the particle jets originated by the b-quark pair from the Higgs boson decay are collimated and new and sophisticated techniques must be applied to separately treat the cases where two b-tagged jets can be individually reconstructed and where the particle boost is such that the two very collimated b-jets are merged in one large jet.

The presented analysis adopts techniques for event pre-selection and b-jet tagging well established from past studies of the ATLAS community working on Higgs boson decays into b-quarks, while the most original part of this thesis is the study of a new technique for treating the case when the Higgs boson is reconstructed from a single large jet. This new technique implements a procedure in which the radius of the cone containing the particles to be associated to the candidate jet, normally fixed a priori, is instead dynamically defined, depending on the jet kinematics (transverse momentum, angular separation from neighboring jets, etc.).

Other relevant parts of the thesis rely on robust and comprehensive studies for defining the signal and control regions in the event phase-space, on a detailed modelling of the shape of the distributions of relevant kinematical variables for the background events, on the optimization for the new analysis of the legacy fit procedure, and on the use of a special framework for measuring the production differential cross-section. All those different studies led to a result which extends the analysis of the Higgs boson decay to a transverse momentum region above 400 GeV never explored before.

Since new physics signals are expected at higher momentum, the results obtained in this analysis are used also to constrain, under certain assumptions, some parameters of the effective field theory considered in this thesis as an extension of the standard model.

Finally, the full analysis framework used in this study of the associated production of the Higgs boson with a boson vector, was applied also to a very similar topology case, the diboson production, where a Z boson decays into a b-quark pair and a Z or W boson decays into leptons. For its completeness and aim of implementing new techniques or improving legacy ones, this thesis may be considered as a solid basis for future developments in the study of the Higgs boson properties with good prospects in the search for new physics signals.

Pisa, Italy
May 2022

Fabrizio Scuri
Paolo Francavilla

Abstract

After the Higgs discovery at LHC in 2012, most of ATLAS analyses are focusing on precise measurements of Higgs kinematic properties and on the search of new decay modes sensitive to physics Beyond the standard model (BSM). Among all the processes, the Higgs boson decay to b-quarks is particularly interesting, thanks to its branching ratio of about 58% in the Standard Model (SM). The observation of this decay at the LHC has been obtained only recently, after 7 years from the Higgs boson discovery, because this channel is affected by large backgrounds arising from multi-jet production that make a real challenge to trigger and extract the signal. The best sensitivity is presently obtained by studying the associated Higgs boson production with a vector boson (W or Z) decaying leptonically.

The analysis described in this thesis is the search of the Higgs boson, decaying into $b\bar{b}$ pair, in the associated production with a vector boson, in the extreme Higgs boson transverse momentum region where the Higgs boson is reconstructed using the large-R jet technique. The use of the large-R jets allows to add a part of the phase space unexplored so far, which is particularly sensitive to possible new physics.

The analysed data have been collected at LHC by the ATLAS detector between 2015 and 2018 at a centre-of-mass energy of $\sqrt{s} = 13$ TeV. The same dataset has been used to perform the differential $pp \rightarrow ZH$ and $pp \rightarrow WH$ cross-section measurements used to extract the information on the Higgs couplings and to put limits on BSM effects.

Furthermore, the analysis has been re-used to perform a cross-section measurement of the diboson ZZ and WZ processes because the diboson and the Higgs processes have a similar topology. For the first time, the $ZZ(b\bar{b})$ and $WZ(b\bar{b})$ cross-sections are measured at $\sqrt{s} = 13$ TeV and the observed cross-section measurements are consistent with the Standard Model predictions.

Preface

The Standard Model (SM) of elementary particles is a theory that describes three of the four fundamental interactions in Nature: weak, strong and electromagnetic interactions. Presently, only the gravitational force is not well described by the SM. Moreover, the SM fails to provide explanations for certain observed phenomena as the presence of almost only matter in the Universe even if antimatter was originally produced with an equal amount, and the presence of dark matter. These and other open questions suggest that the SM can not be considered a complete theory. Several models for physics Beyond the Standard Model (BSM) were developed during the years, and many of them are investigated at the Large Hadron Collider (LHC). The LHC is the world's largest particle accelerator operating at the CERN laboratory that makes collisions at the very high centre-of-mass energy $\sqrt{s} = 13$ TeV. ATLAS is one of the two multi-purpose detectors placed on the LHC ring whose physics program can be divided into two areas: tests the SM predictions and searches for new hints of BSM physics. The analysis described in this thesis uses data collected by the ATLAS detector and both aspects of the ATLAS physics program are explored.

The latest missing piece of the SM has been discovered in July 2012 by the ATLAS [1] and the CMS [2] Collaborations. The discovery of the Higgs boson has been the most recent step forward in experimental particle physics and has opened a whole new sector of studies at the LHC. Since the discovery, the ATLAS and CMS experiments have precisely measured the Higgs boson properties, like its mass, spin-parity, as well as the production, the decay rates and the coupling to bosons and fermions. The dominant decay of the Higgs boson is into pairs of b-quarks with a branching ratio of 58% for a Higgs boson mass $m_H = 125$ GeV. The first evidence of this decay mode has been obtained only after 5 years from the Higgs boson discovery because the large background from multi-jet production makes the search in the dominant gluon-gluon fusion production mode very challenging at the hadron colliders. The best sensitivity is presently obtained by studying the Higgs boson production in association with a vector boson V, which can be either a Z or a W boson, even if this production mode has a cross-section more than one order of magnitude lower than the gluon-gluon fusion. The request of a leptonic decay of the vector boson V enables efficient triggering and significant reduction of the multi-jet background.

Furthermore, the associated production VH is an interesting channel for the study of BSM physics which could change the H-VV coupling structure enhancing this production mode at high energy scale.

The main analysis described in this thesis is on the search of the Higgs boson, decaying into $b\bar{b}$ pair, in the associated production with a vector boson V, in the extreme Higgs boson transverse momentum region, where the Higgs boson is reconstructed using the large-R jet technique. This analysis is new in the ATLAS Collaboration and it is the next natural evolution of the $VH(b\bar{b})$ analysis in which the Higgs decay products are reconstructed using two separate b-jets. The large-R jet technique was proposed by Butterworth et al. [3] almost 10 years ago, it has been already used in searches for BSM physics, but not yet to measure the Higgs production cross-section in association with a vector boson. The use of the large-R jets would allow to add a part of the phase space unexplored so far which is particularly sensitive to possible new physics. The limits to BSM effects are set re-interpreting the so-called Simplified Template Cross-Section (STXS) framework. The STXS framework facilitates the measurement of the differential $pp \rightarrow VH$ cross-section which has been used to extract information on the Higgs couplings and to put limits on BSM effects.

Structure of the Thesis and Personal Contribution

The work presented in this thesis is the outcome of a joint effort among all the collaborators of the $VH(H \rightarrow b\bar{b})$ analyses. The level of details in the chapters reflects my specific contribution to the analysis. The content of this thesis is organized as follows.

Chapter 1 gives an overview of the SM with a focus on the importance of the Higgs mechanism. Moreover, this chapter introduces the reader to the experimental and historical context in which the analysis covered by this thesis is carried out. There is a summary of the previous results published by the ATLAS experiment because my work has started contributing to the $VH(H \rightarrow b\bar{b})$ analysis in which the Higgs decay products are reconstructed with two separate b-jets testing the event selection and the fit procedures. I joined the analysis team in time to be one of the main contributors in performing the statistical tests to obtain the observation of the $H \rightarrow b\bar{b}$ decay mode and the VH production mode in 2018. Moreover, I also produced a good part of the plots and tables in Ref. [4]. The latest results of the $VH(H \rightarrow b\bar{b})$ analysis with two separate b-jets are reported in the last part of Chap. 1.

Chapter 2 describes the Large Hadron Collider accelerator complex, its design and the performance. The description of the ATLAS detector, with its main sub-systems, is also reported.

Chapter 3 is dedicated to the data sample and trigger strategy adopted by the analysis to select the events. This chapter also contains the description of the simulated samples used for the expected signal and background distributions in the different phase space regions explored in the analysis.

Chapter 4 explains the experimental procedures employed in the reconstruction and identification of the physics objects used in the analysis. The analysis team developed an analysis framework to share common definitions for the physics objects. I am responsible for the common E_T^{miss} definition, acting as liaison with the dedicated ATLAS performance group for the analysis team.

Chapter 5 summarizes the list of all the objects and the specific requirements used in the $VH(b\bar{b})$ analysis described in this thesis. In addition, this chapter describes the event selection and categorization of the analysis. I have contributed to the definition of the event selection paying particular attention to the channel aiming at selecting $ZH \rightarrow \nu\nu b\bar{b}$ events. My work was mainly focused on the estimate of the multi-jet background and on the event categorization according to the jet multiplicity.

Chapter 6 shows the source of systematic uncertainties that affect the analysis. A particular attention is given to the modelling uncertainties. I personally extracted the modelling systematic uncertainties coming from the single-top and top-pair processes which are two important backgrounds of the analysis.

Chapter 7 is dedicated to the statistical procedure adopted to extract the analysis results. The chapter is divided into two main parts: the first part covers the fundamental principles of the statistical approach, while the second part describes the specific tools applied to validate and to understand the fit results.

Chapter 8 shows the results obtained by the $VH(b\bar{b})$ analysis using the full Run 2 dataset. I contributed to the definition of the statistical analysis and the procedure to extract the final results. I also documented, together with other three analysers, the entire analysis in the internal supporting note on which the published paper was based. In the second part of the chapter, the STXS measurements are reported together with the limits on BSM effects through the Effective Field Theory (EFT) formalism.

Chapter 9 is fully dedicated to a spin-off of the $VH(b\bar{b})$ analysis focused on high-energy events. The $VH(b\bar{b})$ analysis is re-used to perform cross-section measurements of the diboson ZZ and WZ processes because the Higgs and the diboson processes have a similar topology. A simultaneous fit is performed to extract the VH and VZ cross-sections. This is the first measurement of the $ZZ(b\bar{b})$ and $WZ(b\bar{b})$ cross-sections at $\sqrt{s} = 13$ TeV. The observed cross-section measurements are consistent with the SM predictions and they can be used to put limits on BSM effects. As the sole analyser, the gained experience allowed me to test the SM sector by doing sensitivity studies, producing and testing the diboson simulated samples split into the exclusive phase space regions, and adapting the statistical framework for the diboson cross-section measurement. In addition, I personally computed the theoretical predictions and their uncertainties on the cross-section for the diboson measurements.

Finally, Chap. 10 outlines the general conclusions of the work presented in this thesis, with an outlook on the future prospects for the $VH(b\bar{b})$ analysis.

In parallel to my activity in the analysis of the ATLAS data, during my Ph.D. I continued the collaboration with the community of the central section of the ATLAS hadronic calorimeter, TileCal. Appendix B describes the TileCal layout and the calibration methods used to monitor the TileCal response stability. Special attention is dedicated to the Laser calibration system which monitors the response variation of the photomultipliers reading out TileCal. I personally contributed to the development of a new algorithm used to study the stability of the photomultiplier response. Moreover, I studied the photomultiplier response against ageing to understand if the full sample of photomultipliers installed at the beginning of the ATLAS detector operation can be used until the completion of the High-Luminosity LHC. The studies of the photomultiplier response against ageing are also reported in Appendix B.

Pisa, Italy Giulia Di Gregorio

References

1. ATLAS Collaboration (2012) Observation of a new particle in the search for the Standard Model Higgs boson with the ATLAS detector at the LHC. Phys Lett B 716:1. https://doi.org/10.1016/j.physletb.2012.08.020. arXiv:1207.7214 [hep-ex]
2. CMS Collaboration (2012) Observation of a new boson at a mass of 125 GeV with the CMS experiment at the LHC. Phys Lett B 716:30. https://doi.org/10.1016/j.physletb.2012.08.021. arXiv:1207.7235 [hep-ex]
3. Butterworth JM, Davison AR, Rubin M, Salam GP (2008) Jet substructure as a new Higgs search channel at the LHC. Phys Rev Lett 100:242001. https://doi.org/10.1103/PhysRevLett.100.242001. arXiv:0802.2470 [hep-ph] (cit. on p. xii)
4. ATLAS Collaboration (2018) Observation of $H \rightarrow b\bar{b}$ decays and VH production with the ATLAS detector. Phys Lett B 786:59. https://doi.org/10.1016/j.physletb.2018.09.013. arXiv:1808.08238 [hep-ex]

Acknowledgments

I would like to thank my supervisors, Fabrizio Scuri and Paolo Francavilla, for their guidance in this work. Thanks for the confidence you had in my work and the freedom you gave me to choose what I want to do.

A special thank to Valerio Dao for always believing in me from the first moment (or maybe from the first rush). I appreciate how you helped me entering new sectors of the analysis and the fact that you have always been available to share new ideas and to follow my progress. I really enjoy working with you and I hope I will have the opportunity to continue our collaboration.

I am thankful to the ATLAS Pisa group for the support and for giving me the chance to participate in many interesting conferences and physics schools and to spend one year at CERN, which helped me to get really involved in the analysis. Many thanks to Milène with whom I had the chance not only to work together and share ideas but also I had a lot of fun during the coffee breaks and the evening runs. I should also thank you for the patience you showed in replying to all my questions and doubts.

Further, I would like to thank the $VH(b\bar{b})$ group and in particular Hannah Arnold, for finding always the time to give me hints and to discuss physics. Our discussions allowed me to gain a broader overview of the analysis. I thank you also for the very welcoming atmosphere that you and Valerio created in the $VH(b\bar{b})$ *boosted* analysis.

I would like to thank also the TileCal group for giving me the chance to work in such an amazing environment. A special mention to Henric Wilkens who always supported my ideas and interests.

Thanks to all my colleagues and friends with whom I shared the past years in Pisa. In particular, I would like to mention Claudio Bonanno and Giulia Pagano: even if we were separated in the last period, I really enjoyed to share this crazy experience with you.

I would like to acknowledge my parents for their endless support and curiosity in my research field. You were always ready to encourage me over the years even if my choices brought me far from home. I would like to thank also my brother for supporting me in his own way and, with his girlfriend, giving me the chance to experience being an aunt. Thanks also to my grandparents for always welcoming me home with love and warmth.

Thank you Denis for always being by my side: you have never made me feel alone and you have always shown me how much you believe in me. I also want to thank you because you made me rediscover that curiosity I had lost for a while. Finally, I want to thank you because, even if you do not feel like it, you always protect me and calm me down.

Last but not least I would like to thank myself. This Ph.D. was a roller coaster of emotions. At the beginning, I was a bit scared and sometimes I felt like I was not up to the task. But, who cares?! I finished it and I am proud. Giulia out.

Contents

Chapter 1
Theoretical Aspects and Analysis Overview

After a general overview on the Standard Model with focus on the Higgs mechanism, this chapter illustrates the state of the art of the $VH(b\bar{b})$ analysis. The last sections introduce the framework adopted to extrapolate the information on the physics Beyond the Standard Model from the VH cross-section measurement.

1.1 The Standard Model

The Standard Model (SM) of elementary particles [1–3] is a quantum field theory that describes three of the four fundamental interactions in Nature: strong, weak and electromagnetic interactions. Only the gravitational force, described by Einstein's general relativity, is not described by the Standard Model. Moreover, the gravitational force between two individual particles is extremely small and can be neglected in the discussion of particle interactions.

In the Standard Model the elementary constituents of matter are fermions of spin 1/2 and the interactions are mediated by the exchange of spin 1 particles called bosons. Additionally, the field of the scalar (spin 0) Higgs boson generates their masses. The fermions are classified in leptons and quarks and each fermion has its corresponding anti-particle. Fermions are also divided in three generations of increasing mass, both for leptons and quarks, as shown in Fig. 1.1. The leptons are further categorized into charged leptons (e^-, μ^-, τ^-), which interact weakly and electromagnetically, and neutral leptons called neutrinos (ν_e, ν_μ, ν_τ), which interact only through the weak interactions. The quarks exist in six flavours, up (u), down, (d), charm (c), strange (s), top (t) and bottom (b). They have a fractional charge Q in unit of the positron charge which is +2/3 for up, charm and top quarks, and –1/3 for down, strange and bottom quarks. They can interact strongly, weakly and electromagnetically. Quarks carry a *color charge*, which comes in three flavours, red (r), green (g) and blue (b). In Nature only colorless particles are observed and free quarks have never been detected. The absence of free quarks is explained by the *color confinement* where the coloured objects are always confined to uncoloured

G. Di Gregorio, *A Beauty-ful Boson*, Springer Theses,
https://doi.org/10.1007/978-3-031-20013-7_1

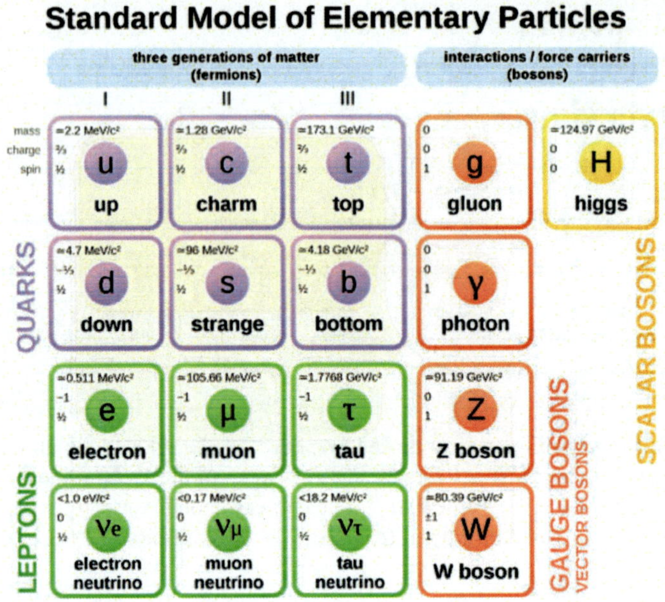

Fig. 1.1 Summary of the Standard Model elementary particles and their properties. The twelve fermions and five fundamental bosons (the eight gluons and the two W bosons are shown in a single g and W box respectively) are shown

particles called *hadrons* and cannot propagate as free particles. Hadrons composed of three quarks are classified as half-integer spin *baryons*, while hadrons composed of a quark anti-quark pair are integer spin particles called *mesons*. The theory describing the strong interactions between quarks is called Quantum Chromodynamics (QCD). Recently particles with four and five quarks, called tetraquarks or pentaquarks, have been observed. The finding will help physicists better understand the complex ways in which quarks bind themselves together into composite particles.

In the Standard Model there are twelve spin 1 bosons: one photon which mediates the electromagnetic force between electrically charged particles, three bosons, W^+, W^- and Z which mediate the weak force, and eight gluons which are the mediators of the strong interactions.

The Standard Model is a relativistic quantum field theory in which the Lagrangian controls the dynamics and the kinematics of theory. Particles are associated to quantum fields depending on the space-time coordinates. The Standard Model Lagrangian is invariant under the local gauge symmetry[1] group $SU(3)_C \times SU(2)_L \times U(1)_Y$. The suffix "C" stands for the conserved charged of the strong interaction referred to as *colour*, the suffix "L" indicates that the vector bosons only couple to left handed

---

[1] A local gauge transformation implies that the considered fields vary differently at any point in space-time.

component of the fermions, while the suffix "Y" indicates the conserved quantity known as *hypercharge*. The symmetries and the interactions of the SM are briefly discussed in the following.

The Lagrangian of a free spin 1/2 particle is:

$$\mathcal{L} = \bar{\psi}\left(i\gamma^{\mu}\partial_{\mu} - m\right)\psi \tag{1.1}$$

where m is the mass of the particle, ψ is the spinor of the particle, γ^{μ} are the Dirac gamma matrices, $\partial_{\mu} = \partial/\partial x_{\mu}$ is the derivative and $\bar{\psi} = \psi^{\dagger}\gamma^{0}$ is the Dirac adjont of the spinor. Under the $U(1)$ local space transformation, the spinor field transforms as:

$$\psi(x) \rightarrow \psi'(x) = \exp(ie\theta(x))\,\psi(x) \tag{1.2}$$

where $\theta(x)$ is a generic function which represents the local phase and e is the elementary charge. With this transformation, the Lagrangian of the free particle becomes:

$$\mathcal{L} \rightarrow \mathcal{L}' = \mathcal{L} - e\bar{\psi}\gamma^{\mu}(\partial_{\mu}\theta(x))\psi \tag{1.3}$$

Hence, as it stands, the free-particle Lagrangian is not invariant under $U(1)$ local phase transformations. To restore the gauge invariance, one can replace the derivative ∂_{μ} with the covariant derivative D_{μ}:

$$\partial_{\mu} \rightarrow D_{\mu} = \partial_{\mu} + ieA_{\mu} \tag{1.4}$$

where A_{μ} is a new field. The desired cancellation of the unwanted term $e\bar{\psi}\gamma^{\mu}(\partial_{\mu}\theta(x))\psi$ is provided by the new field A_{μ} which transforms as:

$$A_{\mu} \rightarrow A'_{\mu} = A_{\mu} - \partial_{\mu}\theta(x) \tag{1.5}$$

The field A_{μ} is interpreted as the gauge field of the electromagnetic interaction. Hence the Quantum Electrodynamics (QED) Lagrangian which describes the fields for the electron, the photon and the interactions between them can be written as:

$$\mathcal{L}_{QED} = \bar{\psi}\left(i\gamma^{\mu}D_{\mu} - m\right)\psi - \frac{1}{4}F_{\mu\nu}F^{\mu\nu} \tag{1.6}$$

where $F_{\mu\nu} = \partial_{\mu}A_{\nu} - \partial_{\nu}A_{\mu}$ and $F_{\mu\nu}F^{\mu\nu}$ is the kinetic energy term of the photon.

While QED is invariant under local $U(1)$ transformation, the gauge symmetry of the QCD is $SU(3)$ which is a non-abelian group.[2] Due to $SU(3)$ gauge symmetry of QCD, the spinor field transforms as:

$$\psi_{A}(x) \rightarrow \psi'_{A}(x) = \exp(i\alpha_{s}\vec{\lambda}\cdot\vec{\beta}(x))\,\psi_{A}(x) \tag{1.7}$$

[2] A non-abelian gauge symmetry group is a gauge set of transformations which do not obey to the commutative law.

where ψ_A is the spinor describing a quark carrying colour A, α_s is the strong coupling constant, $\overrightarrow{\lambda} = (\lambda_1, \lambda_2, ..., \lambda_8)$ are the eight Gell-Mann matrices and $\overrightarrow{\beta}(x) = (\alpha_1(x), \alpha_2(x), ..., \alpha_8(x))$ are the eight angles. The $SU(3)$ gauge invariant QCD Lagrangian is:

$$\mathcal{L}_{QCD} = -\frac{1}{4} G^A_{\alpha\beta} G^{\alpha\beta}_A + \sum_{flavours} \bar{\psi}_A (i\gamma^\mu D_\mu - m)_{AB} \psi_B \tag{1.8}$$

where D_μ is the appropriate SU(3) covariant derivative and $-\frac{1}{4} G^A_{\alpha\beta} G^{\alpha\beta}_A$ is the kinematic term of the eight massless QCD mediators called gluons. The $G^A_{\alpha\beta}$ is the gluonic field tensor which is found from the gluon field A^A_α:

$$G^A_{\alpha\beta} = \partial_\alpha A^A_\beta - \partial_\beta A^A_\alpha - g_S f^{ABC} A^B_\alpha A^C_\beta \tag{1.9}$$

where the indices A, B, C run over the colour degrees of freedom of the gluon fields, the coupling constant g_S determines the strength of the interaction between coloured partons and, f^{ABC} are the structure constants of the $SU(3)$ colour group. In QCD the single charge of QED is replaced by the three color charges. Only quarks, which are particles with non-zero colour charge, couple to gluons.

So far it has been shown that QED and QCD are associated with $U(1)$ and $SU(3)$ local gauge symmetry, respectively. The charge weak interaction is associated with the invariance under $SU(2)$ local phase transformations defined as:

$$\psi(x) \rightarrow \psi'(x) = \exp\left[\frac{g_W}{2} \overrightarrow{\alpha}(x) \cdot \overrightarrow{\sigma}\right] \psi(x) \tag{1.10}$$

where $\overrightarrow{\sigma}$ are the three Pauli spin matrices that are generators of the $SU(2)$ group, $\overrightarrow{\alpha}(x) = (\alpha_1(x), \alpha_2(x), \alpha_3(x))$ are the three functions which specify the local phase at each point in space-time and g_W is the gauge coupling constant. The required local gauge invariance can be only satisfied by the introduction of three gauge fields, $W^{(k)}_\mu$ with $k = 1, 2, 3$ which are the analogous of A_μ in QED. These gauge fields correspond to three gauge bosons $W^{(1)}$, $W^{(2)}$ and $W^{(3)}$. The charged currents are expressed as a linear combination of $W^{(1)}_\mu$ and $W^{(2)}_\mu$:

$$W^\pm_\mu = \frac{1}{\sqrt{2}} (W^{(1)}_\mu \pm W^{(2)}_\mu) \tag{1.11}$$

The wave function $\psi(x)$ is written in terms of two components, because the generators of the $SU(2)$ gauge transformations are the 2×2 Pauli matrices, and it is called weak isospin doublet. The weak isospin doublet contains flavours differing by one unit of electric charge, i.e. the neutrino-electron isospin doublet:

$$\psi_L(x) = \begin{pmatrix} \nu_e(x) \\ e^-(x) \end{pmatrix}_L \tag{1.12}$$

In this example, ν_e and e^- have a total weak isospin $I_W = \frac{1}{2}$ with third component of the weak isospin $I_W^{(3)}(\nu_e) = +\frac{1}{2}$ and $I_W^{(3)}(e^-) = -\frac{1}{2}$. Measurements show that the weak charged-current interaction couples only to left-handed particles and right-handed anti-particles, so the gauge $SU(2)$ transformation affects only left-handed particles and right-handed anti-particles. The right-handed particles and the left-handed anti-particles have a null weak isospin $I_W = 0$ and they are not affected by the $SU(2)$ local gauge transformation. For this reason, the symmetry group of the weak interaction is referred to as $SU(2)_L$ because the doublets are composed only of left-handed particles and right-handed anti-particles. The right-handed particles are placed in weak isospin singlets with $I_W = I_W^{(3)} = 0$, e.g:

$$e_R^-, \quad u_R, \quad d_R \tag{1.13}$$

Experiments show that the Z boson couples to both left-handed and right-handed states so it can not correspond to the $W^{(3)}$ of the $SU(2)_L$ local gauge symmetry. Also the photon couples left-handed and right-handed particles. In the electroweak model of Glashow, Salam and Weinberg the $U(1)$ gauge symmetry of QED is replaced with a new $U(1)_Y$ local gauge symmetry of weak hypercharge Y. The weak hypercharge is defined as:

$$Y = 2Q - 2I_W^{(3)} \tag{1.14}$$

where Q is the electromagnetic charge of the fermion and $I_W^{(3)}$ is the third component of the the weak isospin. The combination of $SU(2)_L \times U(1)_Y$ generates four gauge fields that describe the electroweak interactions. The $U(1)_Y$ local gauge symmetry gives rise to a new field B_μ. The physics fields A_μ, Z_μ, W_μ^+, W_μ^- corresponding to the γ, Z and $W^\pm$ bosons are a linear combination of B, $W^{(1)}$, $W^{(2)}$, $W^{(3)}$. $W_\mu^\pm$ and Z_μ are the fields associated to the weak bosons (Z, $W^\pm$), while A_μ is the field representing the photon. The weak neutral gauge field and the photon field are expressed as a linear combination of B_μ and $W_\mu^{(3)}$:

$$\begin{pmatrix} Z_\mu \\ A_\mu \end{pmatrix} = \begin{pmatrix} \cos\theta_W & -\sin\theta_W \\ \sin\theta_W & \cos\theta_W \end{pmatrix} \begin{pmatrix} W_\mu^3 \\ B_\mu \end{pmatrix} \tag{1.15}$$

where θ_W is the weak mixing Weinberg angle defined as:

$$\frac{g_W}{\sin\theta_W} = \frac{g'}{\cos\theta_W} = e \tag{1.16}$$

where e is the electron charge, g_W is the coupling constant of the $SU(2)_L$ local gauge transformation and g' is the coupling constant of the $U(1)_Y$ local gauge transformation. The SM described so far does not include the mass of the bosons and fermions. The masses of the particles are included in the SM with the Higgs mechanism.

1.2 The Higgs Mechanism

Constructing the Standard Model on symmetries, one of the striking conclusions is that all the particles in the model should be massless because an explicit mass term in the Lagrangian both for bosons and fermions would violate the local invariance. On the contrary, experimental measurements have shown that particles, as the electroweak gauge bosons, are massive. The mass generation of the electroweak gauge bosons and of the fermions is included through the Brout-Englert-Higgs mechanism [4–6], also called Higgs mechanism, with the spontaneous symmetry breaking. This mechanism introduces a scalar Higgs term to the SM Lagrangian:

$$\mathcal{L}_{Higgs} = (D^{\mu}\phi)^{\dagger}\left(D_{\mu}\phi\right) - V(\phi) \tag{1.17}$$

where ϕ is a complex scalar field, $V(\phi)$ is the potential of the Higgs scalar field and D_{μ} is the covariant derivative of a complex scalar field. The minimal Higgs model consists of two complex scalar fields:

$$\phi = \begin{pmatrix} \phi^{+} \\ \phi^{0} \end{pmatrix} = \frac{1}{\sqrt{2}}\begin{pmatrix} \phi_1 + i\phi_2 \\ \phi_3 + i\phi_4 \end{pmatrix} \tag{1.18}$$

The covariant derivative of ϕ is:

$$D_{\mu}\phi = \left(\partial_{\mu} + i\frac{g_W}{2}\overrightarrow{\sigma}\cdot\overrightarrow{W_{\mu}} + i\frac{g'}{2}B_{\mu}\right)\phi \tag{1.19}$$

where $\overrightarrow{W_{\mu}} = (W_{\mu}^{(1)}, W_{\mu}^{(2)}, W_{\mu}^{(3)})$ and B_{μ} are the $SU(2)_L$ and $U(1)_Y$ gauge bosons introduced in the previous section. The Higgs potential $V(\phi)$ is:

$$V(\phi) = \mu^2\phi^{\dagger}\phi + \lambda\left(\phi^{\dagger}\phi\right)^2 \tag{1.20}$$

where μ and λ are scalar constants and, in particular, λ describes the quadratic self-interaction among the scalar fields. The spontaneous symmetry breaking is based on the non-invariance of the vacuum state with respect to the $SU(2)$ symmetry. When μ^2 and λ are both positive, the minimum of the potential is found in the unique configuration $\phi = 0$. If $\mu^2 < 0$ and $\lambda > 0$, the minimum of the potential $V(\phi)$ is described by an infinite number of solutions satisfying:

$$\phi^{\dagger}\phi = \frac{1}{2}\left(\phi_1^2 + \phi_2^2 + \phi_3^2 + \phi_4^2\right) = \frac{v^2}{2} = -\frac{\mu^2}{2\lambda} \tag{1.21}$$

where v is the vacuum expectation value. Figure 1.2 shows the potential $V(\phi)$ for the values $\mu^2 > 0$ and $\mu^2 < 0$. The symmetry is spontaneously broken choosing the system to fall into one of the multiple ground states, for example:

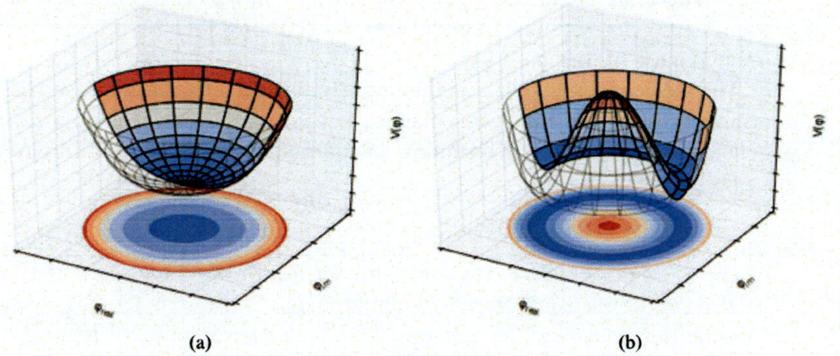

Fig. 1.2 Higgs potential $V(\phi)$ for a complex scalar field for $\mu^2 > 0$ (a) and $\mu^2 < 0$ (b)

$$\phi_0 =< 0|\phi|0 >= \frac{1}{\sqrt{2}}\begin{pmatrix} 0 \\ v \end{pmatrix} \tag{1.22}$$

Expanding around the ground state, the Higgs field itself is given by:

$$\phi(x) = \frac{1}{\sqrt{2}}\begin{pmatrix} 0 \\ v + h(x) \end{pmatrix} \tag{1.23}$$

where $h(x)$ is the real scalar field corresponding to the Higgs boson.

By substituting the Higgs field in the Higgs Lagrangian (Eq. 1.17), the mass terms of the gauge bosons can be identified in the Lagrangian. The values obtained are:

$$m_W = \frac{1}{2}vg_W, \quad m_Z = \frac{1}{2}v\sqrt{g_W^2 + g'^2}, \quad m_A = 0 \tag{1.24}$$

The mass of the Higgs boson particle derived from the expansion of the scalar potential is:

$$m_H = v\sqrt{2\lambda} \tag{1.25}$$

and it is the free parameter of the model given its dependence on λ.

The other terms in the expansion of the covariant derivative define the interaction vertices between the Higgs scalar field $h(x)$ and the gauge bosons:

$$\mathcal{L}_{VH} = \mathcal{L}_{VVH} + \mathcal{L}_{VVHH} \tag{1.26}$$

where

$$\begin{aligned} \mathcal{L}_{VVH} &= \frac{2m_W^2}{v}W_\mu^+ W^{-\mu} h(x) + \frac{m_Z^2}{v} Z^\mu Z_\mu h(x) \\ \mathcal{L}_{VVHH} &= \frac{m_W^2}{v^2}W_\mu^+ W^{-\mu} h(x)h(x) + \frac{m_Z^2}{v^2} Z^\mu Z_\mu h(x)h(x) \end{aligned} \tag{1.27}$$

The vertices are composed by one (or two in $\mathcal{L}_{VVHH}$) Higgs bosons and a couple of massive gauge bosons. In both cases the coupling strength is proportional to the squared mass of the gauge boson.

The mechanism used to explain how fermions acquire mass is based on the same principles of the vector boson masses, but the formalism is slightly different. To generate the masses of the fermions, an additional Yukawa term is introduced in the Lagrangian:

$$\mathcal{L}_{Yukawa} = -g_f \left(\bar{\Psi}_L \phi \Psi_R + \bar{\Psi}_R \phi^\dagger \Psi_l \right) \tag{1.28}$$

where g_f is the Yukawa coupling term for a fermion f, $\Psi_{L(R)}$ is the left (right) handed fermion isospin doublet (singlet) and ϕ is the complex Higgs scalar field. Considering the case of the electron, the Yukawa term can be written as:

$$\begin{aligned}\mathcal{L}_e &= -\frac{g_e}{\sqrt{2}} \left[(\bar{\nu}_e, \bar{e})_L \begin{pmatrix} 0 \\ v+h \end{pmatrix} e_R + \bar{e}_R \, (0, v+h) \begin{pmatrix} \nu_e \\ e \end{pmatrix}_L \right] \\ &= -\frac{g_e v}{\sqrt{2}} (\bar{e}_L e_R + \bar{e}_R e_L) - \frac{g_e h}{\sqrt{2}} (\bar{e}_L e_R + \bar{e}_R e_L)\end{aligned} \tag{1.29}$$

The g_e Yukawa coupling is not predicted by the Higgs mechanism but it is determined interpreting the term $\frac{g_e v}{\sqrt{2}}$ as the electron mass ($\frac{g_e v}{\sqrt{2}} = m_e$). The second term gives rise to a coupling between the electron and the Higgs boson itself.

The above formalism gives only masses to the lower component of the doublet so it can only explain the mass of the charged leptons and the down-type quarks. The mechanism to explain the mass of the up-type quarks requires the introduction of the Hermitian conjugate of the Higgs scalar field ϕ^C:

$$\phi^c = -i\sigma_2 \phi^* = -\frac{1}{\sqrt{2}} \begin{pmatrix} v+h \\ 0 \end{pmatrix} \tag{1.30}$$

The conjugate doublet transforms in the same way as the doublet ϕ and it introduces the masses of the up-type quarks.

1.3 The Higgs Boson at LHC

The Higgs boson is a scalar particle associated to the Higgs boson field. In the Standard Model the Higgs boson is predicted to be a neutral CP-even scalar ($J^{PC} = 0^{++}$). Its mass m_H is a free parameter of the theory. When the mass of the Higgs boson is fixed, its couplings are well known, the productions rates and the partial widths can be calculated.

1.3.1 Higgs Boson Production Modes

The main Higgs boson production modes at LHC are:

(a) *gluon-gluon fusion* (ggF): at high center-of-mass energy, the gluon-gluon fusion $pp \rightarrow gg \rightarrow H$ is the Higgs boson production with the largest cross-section.[3] This production is mediated by the exchange of a virtual heavy quark (top or bottom) loop. The contribution from lighter quarks propagating in the loop are suppressed proportionally to m_q^2.

(b) *vector boson fusion* (VBF): it is the process with the second largest cross-section. Two W or Z bosons produced from colliding quarks interact to originate the Higgs boson. The scattered quarks give rise to two hard jets in the forward and backward regions. The jets are the characteristic signature of the process used in the analyses to exploit this production mode.

(c) *associated production with a vector boson* (VH): in this channel the Higgs boson is produced in association with a W or a Z boson. As for the VBF case, this channel is driven by the interaction of quarks which produce the vector boson V (with $V = W$ or Z) that emits the Higgs boson. The ZH production also has the contribution from the two gluons initial state. The presence of the vector boson in the finals state is used experimentally to better identify the events as well as to reduce the contribution of background events.

(d) *associated production with a pair of heavy quarks* ($t\bar{t}H$, $b\bar{b}H$): two colliding gluons emit quark - anti-quark pair in which the quark can be the top or the bottom quark. One quark from one gluon and an anti-quark from the other gluon combine and form a Higgs boson. These production mechanisms have the lowest cross-sections at LHC but they present the opportunity to study the direct coupling of the Higgs boson to fermions.

Figure 1.3 shows the Feynman diagrams for the main Higgs production modes at LHC and Fig. 1.4 shows the cross-sections of the Higgs boson with $m_H = 125$ GeV as a function of the center-of-mass energy $\sqrt{s}$.

Table 1.1 summarizes the Higgs boson production cross-sections for a Higgs boson mass $m_H = 125$ GeV at a center-of-mass energy $\sqrt{s} = 13$ TeV.

1.3.2 Higgs Boson Decays

The Higgs boson has a very short lifetime and it decays into final state with fermions or bosons. The theoretical total decay width of the Higgs boson with $m_H = 125$ GeV is $\Gamma_H = 4$ MeV. The branching ratio (BR) to any individual mode is expressed as

[3] The gluon fusion represents almost 90% of the total Higgs cross-section.

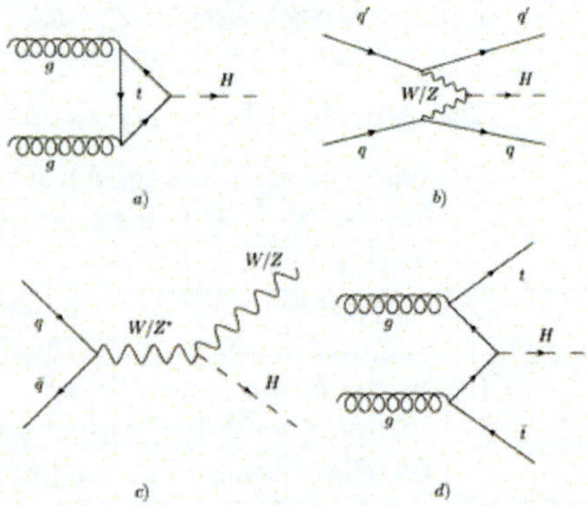

Fig. 1.3 Feynman diagram for the Higgs production: **a** *gluon-gluon fusion*, **b** *vector boson fusion*, **c** *associated production with a vector boson*, **d** *associated production with a pair of top quarks*

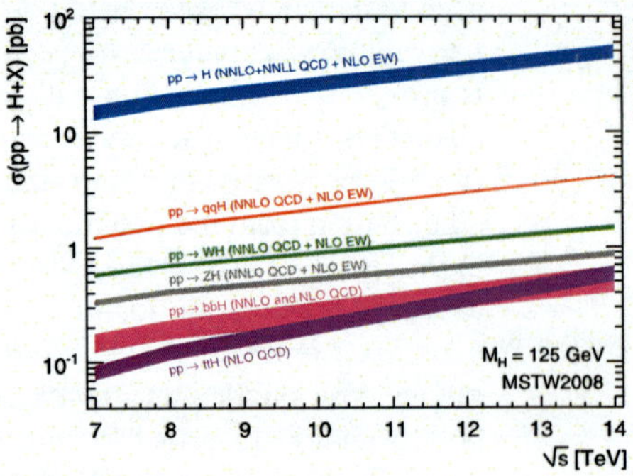

Fig. 1.4 Cross-section production of Higgs boson with $m_H = 125$ GeV as a function of the center-of-mass energy $\sqrt{s}$ [7]

Table 1.1 The Higgs boson production cross-sections for $m_H = 125$ GeV in *p*-*p* collisions at a center-of-mass energy $\sqrt{s} = 13$ TeV [8]

Production mode	σ [pb]
$ggF\ (pp \rightarrow H)$	48.6
$VBF\ (pp \rightarrow qqH)$	3.78
$WH\ (pp \rightarrow WH)$	1.37
$ZH\ (pp \rightarrow ZH)$	0.88
$t\bar{t}H\ (pp \rightarrow t\bar{t}H)$	0.50
$b\bar{b}H\ (pp \rightarrow b\bar{b}H)$	0.48
Total	55.61

the ratio of the partial width to the total width, where the total width is the sum of all possible partial widths:

$$BR(H \rightarrow XX) = \frac{\Gamma(H \rightarrow XX)}{\sum_i \Gamma(H \rightarrow X_i X_i)} \tag{1.31}$$

Figure 1.5 shows the predicted branching ratios of the different decay modes of the Higgs boson as function of its mass. For a Higgs boson with $m_H = 125$ MeV, the

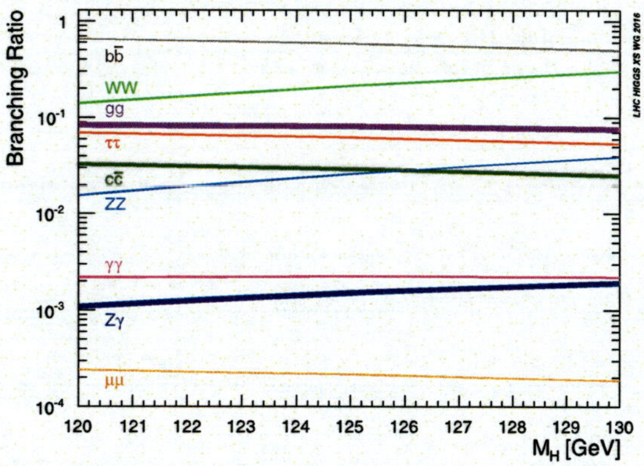

Fig. 1.5 Predicted branching ratios of Higgs boson as a function of its mass [7]

dominant branching ratio is to bottom pairs ($BR \sim 58\%$). This thesis is focused in this decay mode.

The second largest branching ratio is in pairs of W bosons (BR ~22%) with one of the bosons off-shell. The W boson can subsequently decay either into quarks or into a charged lepton and a neutrino. Events with W decays into quark pairs are difficult to distinguish from the QCD background while the W decays into leptons must account for the neutrino missing momentum.

The decay of the Higgs bosons into gluon pairs has the third largest branching fraction but it is not distinguishable from the SM background. For this reason, this decay mode is not studied at LHC.

The next higher BR is in τ lepton pairs. As in $H \rightarrow WW$ case, the main difficulty of this channel comes from the reconstruction of a final state with undetectable neutrinos from the τ lepton decays and the discrimination between the hadron decays of τ leptons and QCD background.

The Higgs to charm pair decay has a low branching ratio ($BR \sim 3$ %) and it suffers from large contamination from QCD process as $g \rightarrow c\bar{c}$ and the additional experimental challenge of tagging hadronic jets from charm quarks.

With a similar branching ratio, the $H \rightarrow ZZ$ decay has a much cleaner experimental signature because of the lepton decay of the Z boson. Even if the production rate of the $H \rightarrow ZZ^* \rightarrow 4l$ decay is extremely low, this channel has small background contributions.

The $H \rightarrow \gamma\gamma$ decay channel has a very low decay rate ($BR \sim 0.2$ %) but it is a clean process as $H \rightarrow ZZ^* \rightarrow 4l$. Since the Higgs boson has no direct coupling to massless particles, the $H \rightarrow \gamma\gamma$ decays is a loop-induced decay with the main contribution from the top quark and W boson. The main background contribution in this channel arises from the SM photon pair production, but the energy and the momentum of the photons are measured with high precision so it is possible to separate the background from the signal with high efficiency. Another rare loop-induced decay mode of interest at LHC is the $H \rightarrow Z\gamma$ production with a branching ratio $BR = 0.15$ %.

Table 1.2 Branching ratios of the Higgs decay for a Higgs boson mass $m_H = 125$ GeV [8]

Decay channel	Branching ratio (%)
$H \to b\bar{b}$	58.2
$H \to WW^*$	21.4
$H \to gg$	8.19
$H \to \tau^+\tau^-$	6.27
$H \to c\bar{c}$	2.89
$H \to ZZ^*$	2.62
$H \to \gamma\gamma$	0.23
$H \to Z\gamma$	0.15
$H \to \mu^+\mu^-$	0.03

The Higgs boson can decay into a μ lepton pair. Despite its very low BR ($BR =$ 0.03%) this decay mode is important because it gives the opportunity to measure the Higgs couplings to the second generation of fermions at LHC.

Table 1.2 summarises the values of the branching ratios for the Higgs boson mass $m_H = 125$ GeV.

1.4 State of the Art of the $VH(b\bar{b})$ Analysis

The discovery of the Higgs boson by ATLAS and CMS Collaborations [9, 10] in 2012 has been the most important recent breakthrough in experimental particle physics and an important test of the Standard Model predictions. After the Higgs boson discovery, many data analyses have focused the attention on its properties. Among all the processes, the study of the Higgs boson decay into b-quarks is particularly important because it is the decay channel with the largest branching ratio. However, large backgrounds from multi-jet production make the search in the dominant gluon-gluon fusion production mode very challenging at hadron colliders. The best sensitivity is presently obtained by studying the Higgs boson produced in association with a vector boson V, with $V = Z$ or $V = W$, even if this production mode has a cross-section which is more than one order of magnitude lower than the gluon-gluon fusion. The leptonic decay of the vector boson V enables efficient triggering and a significant reduction of the multi-jet background. Moreover, this measurement provides the best sensitivity to the WH and ZH production modes and it has the sensitivity to probe some Beyond the Standard Model in effective field theories which could change the $H - VV$ coupling structure [11].

In summer 2017, the first evidence of the Higgs boson decay in $b\bar{b}$ pair was observed by ATLAS [12] and CMS Collaborations [13]. The result was obtained using 2015 and 2016 data, collected at a centre-of-mass energy $\sqrt{s} = 13$ TeV corresponding to an integrated luminosity of 36 fb^{-1} and combined with Run 1 data collected at $\sqrt{s} = 7$ TeV and 8 TeV. In this analysis, called *resolved* analysis, the

Higgs boson has been reconstructed requiring in the final state two b-jets with a radius parameter $R = 0.4$. For a Higgs boson mass of 125 GeV, an excess of events over the expected background from other SM processes was found by the ATLAS Collaboration with an observed significance of 3.5 standard deviations. This result was confirmed by the CMS Collaboration which measured an excess of events over the expected background with an observed significance of 3.8 standard deviations.

Using 2015–2017 data collected at centre-of-mass energy $\sqrt{s} = 13$ TeV, corresponding to a total amount of 79.8 fb^{-1} of integrated luminosity, in summer 2018 ATLAS and CMS Collaborations measured an excess with an observed (expected) significance of 4.9 (4.3) standard deviations and 4.4 (4.2) standard deviations, respectively [14, 15].

The result of the $VH(b\bar{b})$ analysis was combined with other searches for the Standard Model Higgs boson decaying into a $b\bar{b}$ pair, using Run 1 and Run 2 dataset, including the associated production with $t\bar{t}$ pair and the vector boson fusion. For a Higgs boson mass of 125 GeV, the ATLAS Collaboration measured for the $H \to b\bar{b}$ decay a significance of 5.4 standard deviations, to be compared with an expectation of 5.5 standard deviations [14]. The $VH(b\bar{b})$ analysis had the leading sensitivity with a significance of 4.9 standard deviations, while the $t\bar{t}H(b\bar{b})$ and vector boson fusion analyses gave a contribution of 1.9 and 1.5 standard deviations, respectively. The observation of the $H \to b\bar{b}$ decay was confirmed also by the CMS Collaboration [15]. Moreover, the ATLAS Collaboration combined the $VH(b\bar{b})$ result with other Run 2 searches for the Higgs boson in the VH production mode but decaying into either two photons or four leptons via ZZ^* decays. For a Higgs boson mass of 125 GeV, a significance of 5.3 standard deviations was measured [14], above the 5 standard deviations threshold to claim the observation of VH production mode. Most of the significance came from the $VH(b\bar{b})$ analysis (4.9 standard deviations) while the other two channels gave an extra 1.1 standard deviations ($VH(ZZ^*)$ analysis) and 1.9 standard deviations ($VH(\gamma\gamma)$ analysis).

After the observation of the VH production and $H \to b\bar{b}$ decay modes, the same dataset has been re-used in the Simplified Template Cross-Sections (STXS) framework [16, 17]. The STXS framework facilitates the measurement of the differential $pp \to VH$ cross-section and it has been used to extract the information on the Higgs couplings and to put limits on the BSM effects. In autumn 2018 the ATLAS Collaboration announced the measurement of the differential cross-sections of the associated production of the Higgs boson decaying to b-quarks with a vector boson V as a function of the vector boson transverse momentum [18]. All the measurements are in agreement with the Standard Model predictions, and limits are set on parameters of an effective Lagrangian sensitive to modifications of the Higgs boson couplings to the electroweak gauge bosons.

Using the full Run 2 dataset with an integrated luminosity of 139 fb^{-1} the $VH(H \to b\bar{b})$ *resolved* analysis has been updated by the ATLAS Collaboration [19]. An excess of events over the expected background from other SM processes was found by the ATLAS Collaboration with an observed significance of 4.0 and 5.3 standard deviations for a Higgs boson produced in association with a W or a Z boson, respectively. Moreover cross-sections of associated production of a Higgs boson decaying into bottom quark pairs with an electroweak gauge boson decaying into leptons are measured as a function of the gauge boson transverse momentum. The cross-section measurements are all consistent with the Standard Model expectations.

In parallel of the full Run 2 $VH(H \to b\bar{b})$ *resolved* analysis, the $H \to b\bar{b}$ decay mode has been explored in the extreme Higgs boson transverse momentum region reconstructing the Higgs boson with a single large-R jet with $R = 1.0$ [20]. This $VH(H \to b\bar{b})$ analysis designed for the high energy phase space is called *boosted* analysis. Cross-sections of associated production of a Higgs boson decaying into a b quark pair with a V gauge boson are measured in two exclusive V boson transverse momentum regions, 250–400 GeV and above 400 GeV. The region with the V boson transverse momentum above 400 GeV has never been investigated by the *resolved* analysis and it is particularly sensitive to deviations from BSM physics. The main analysis described in this thesis is the $VH(b\bar{b})$ *boosted* analysis with focus on events in the high energy regions.

In the next sections, a brief description of the latest results of *resolved* analysis, the STXS framework and the Effective Field Theory are presented.

1.4.1 Overview of the $VH(b\bar{b})$ Resolved Analysis

The most promising channel to measure the $H \to b\bar{b}$ decay is the $VH(b\bar{b})$ channel. Events are separated into three analysis channels in the so called 0-, 1- and 2-lepton channels, based on the number of charged leptons (electrons or muons) coming from the V boson decay to target the $ZH \to \nu\bar{\nu}b\bar{b}$, $WH \to l\nu b\bar{b}$ and $ZH \to llb\bar{b}$ signatures, respectively. In all channels events are required to have exactly two b-tagged small-R jets with $R = 0.4$ in order to reconstruct the decay products of the Higgs boson. Events are further categorized depending on whether additional, untagged jets are present. In 0- and 1-lepton channels, only events with exactly one untagged jet are considered, while in the 2-lepton channel events with any number of untagged jets are accepted.

To increase the signal-to-background ratio, the three channels are additionally categorized according to the vector boson transverse momentum p_{T}^{V}. In the 0- and 1-lepton channels there are two p_{T}^{V} regions, 150 GeV $< p_{\mathrm{T}}^{V} <$ 250 GeV and p_{T}^{V} $>$ 250 GeV. Due to the stronger background suppression at low p_{T}^{V} in the 2-lepton channel, one additional region has been added for events with 75 GeV $< p_{\mathrm{T}}^{V} <$ 150 GeV.

The dataset used in the analysis has been collected by the ATLAS experiment from 2015 to 2018 with $\sqrt{s} = 13$ TeV and selected using missing transverse momentum ($E_{\mathrm{T}}^{\mathrm{miss}}$) trigger and single lepton triggers. More details on the trigger selection are reported in Sect. 3.1. The offline event selection is performed using physics objects, described in Chap. 4, reconstructed from the detector signal. The definition of the three lepton channels depends on the number of charged leptons in the final states. In the 1-lepton channel, the definition of lepton uses tighter identification and isolation criteria to suppress the multi-jet background.

In 0-lepton channel, events with leptons are rejected. A cut is applied to remove events in a small part of the phase space where the trigger is inefficient. Finally requirements on the angular distance between reconstructed objects are used to suppress the multi-jet background contribution.

In 1-lepton channel, events are required to have one electron or muon. To suppress the multi-jet contribution in the electron sub-channel, an additional selection of $E_{\mathrm{T}}^{\mathrm{miss}} > 30$ GeV is applied.

Events in 2-lepton channel are required to have exactly two leptons with the same flavour. In the muon sub-channel, the leptons are required to have also the opposite charge. To suppress backgrounds with non-resonant leptons, the invariant mass of the lepton pair must be close to the Z boson mass.

Dedicated control regions are defined to evaluate the contributions of the dominant background processes . The control regions are defined using a continuous cut on the angular separation between the two b-tagged jets $\Delta R(\mathbf{b}_1, \mathbf{b}_2)$ as a function of p_{T}^{V}.

To maximise the sensitivity to the Higgs boson signal the analysis deploys a multivariate discriminant, built from variables which describe the kinematics of the selected events. The Boosted Decision Tree (BDT) is the multivariate discriminant used in the analysis. It takes as input kinematic variables that describe the event (p_{T}^{V},$\Delta R(\mathbf{b}_1, \mathbf{b}_2)$ and invariant mass of the two b-tagged jets m_{bb}) and it gives as output a variable called BDT score. The BDT score variable tends to assume values near -1 for background events and values near to +1 for signal events in order to separate signal from background events. A BDT distribution is built for each region of the analysis. The BDT outputs are then combined using a binned maximum likelihood fit to extract the signal strength μ and the background normalisations. The signal strength is defined as:

$$\mu_{VH}^{bb} = \frac{\left(\sigma(VH) \times BR(H \to b\bar{b})\right)_{\text{measured}}}{\left(\sigma(VH) \times BR(H \to b\bar{b})\right)_{\text{expected(SM)}}} \tag{1.32}$$

where σ is the VH cross-section and BR is the branching ratio $H \to b\bar{b}$.

Considering all data collected by the ATLAS Collaboration from 2015 to 2018 at $\sqrt{s} = 13$ TeV, corresponding to an integrated luminosity of 139 fb^{-1}, an excess of events over the expected background from the other SM processes is found with an observed significance of 6.7 standard deviations [19]. The fitted value of the signal strength parameter is $\mu_{VH}^{bb} = 1.02^{+0.18}_{-0.17}$. Moreover the fit is performed measuring separately the WH and ZH production processes. The WH and ZH production modes reject the background-only hypothesis with observed significance of 4.0 and 5.3 standard deviations, respectively. The fitted values of the signal strengths are:

$$\begin{aligned} \mu_{WH}^{bb} &= 0.95^{+0.27}_{-0.25} \\ \mu_{ZH}^{bb} &= 1.08^{+0.25}_{-0.23} \end{aligned} \tag{1.33}$$

with a linear correlation of 2.7% between them.

As a validation, a diboson analysis has been performed in which dedicated BDT distributions are evaluated considering the VZ diboson process as signal and the VH as background. The measurement of the VZ production returns a signal strength of $\mu^{bb}_{VZ} = 0.93^{+0.15}_{-0.14}$, which is in good agreement with the Standard Model prediction.

Another cross-check analysis is the so-called *dijet-mass* analysis in which the distributions of the invariant mass of the two b-tagged jets m_{bb} are used in the fit instead of the BDT output distributions. From the fit to m_{bb}, the value of the signal strength is $\mu^{bb}_{VH} = 1.17^{+0.25}_{-0.23}$ and an observed significance of 5.5 standard deviations has been measured.

1.5 The Simplified Template Cross-Section Measurements

Many successful results have been obtained during Run 1 and the early Run 2, among which the ATLAS $H \rightarrow b\bar{b}$ observation and the VH observation. Such analyses show their results as measured signal strengths and multiplicative coupling modifiers. The Simplified Template Cross-Section (STXS) framework has been developed to provide a natural way to evolve the signal strength measurements. The framework provides a way to perform measurements that are more granular than the signal strength measurements and it allows for an easy combination of the results in different decay channels and different experiments.

The STXS measurements have been performed in mutually exclusive regions of phase space, called STXS bins, and they have a clear advantage with respect to the signal strength measurements. In the signal strength measurement only the signal normalization can be changed and the shape of the kinematic distributions are set to the SM predictions. In the STXS measurement, only the sum of the templates is forced to be identical to the SM prediction while measurements in each bin can show some deviations from the SM. In this regard, the primary goals of the STXS framework are to maximize the sensitivity of the measurements while reducing the theory dependences that must be directly folded into the measurements.

For the STXS measurements, the simulated signal samples are used to create one set of histograms for each STXS bin. Any simulated signal event passing the event selection is assigned to a specific STXS bin and used to fill a specific histogram. Each STXS bin is defined using the generator values of the measured quantities. The quantities obtained at generator level, called truth quantities in the following, used in the VH analysis are:

- Higgs boson rapidity y_H which is required to be lower than 2.5 ($|y_H| < 2.5$) to reduce the extrapolation to region with limited signal acceptance;
- transverse momentum of the vector boson p_T^V: the STXS bins are designed to be coherent with the cuts used in the VH analysis and to isolate regions sensitive to BSM physics;
- number of small-R jets with $R = 0.4$ and $p_T > 30$ GeV: the STXS bins are designed to mitigate the extrapolation among regions with different jet multiplicity.

The criteria used to define the STXS bins are:

- minimizing the dependence on the theoretical uncertainties;
- maximizing the experimental sensitivity;
- isolation of possible BSM effects;
- minimizing the number of bins without loss of experimental sensitivity.

If the residual theoretical dependence into the signal acceptance in each STXS bin is still comparable to the precision measurement, one can split a bin into two or smaller bins. Moreover, the use of standardized STXS bins, which are defined with no information on the Higgs decay modes, makes the STXS framework extremely adaptable for combinations among decay channels, and for comparison among experiments.

The STXS measurements are different from the fully fiducial differential measurements. The latter are explicitly optimized for maximal theory independence by minimizing the acceptance corrections. The minimization of the acceptance corrections is done performing measurements in fiducial volumes which are as close as possible to fiducial volumes measured for a particular Higgs boson decay channel. In contrast, in the STXS framework, simplified fiducial volumes are used and larger acceptance corrections are allowed in order to maximally benefit from the use of standardized event categories. Another difference is that the fiducial measurement is designed to be agnostic to the production mode while the separation in production modes is an essential aspect of the STXS framework.

In the STXS framework, the VH production is referred as the Higgs boson production in association with a V boson decaying leptonically. In the VH STXS study, a so-called *maximum splitting* categorization scheme is proposed to provide the flexibility to convert any categorization scheme of interest. The VH signals are divided into five categories: $qq \rightarrow Z(\nu\nu)H$, $gg \rightarrow Z(\nu\nu)H$, $qq \rightarrow W(l\nu)H$, $qq \rightarrow Z(ll)H$ and $gg \rightarrow Z(ll)H$. All the lepton flavours are considered (including τ-leptons). As anticipated, each category is further divided into a forward Higgs region and a central Higgs region. The former includes events where the Higgs boson rapidity is higher than 2.5, while in the latter the Higgs rapidity is lower than 2.5 (for which the experimental acceptance is negligible). The signals with a central Higgs are further split using the vector boson transverse momentum p_{T}^{V}. Five p_{T}^{V} bins are defined with four cuts at 75, 150, 250 and 400 GeV. In each p_{T}^{V} interval, events are separated according to the number of extra jets. Extra jets are defined as small-R jets with $R = 0.4$ and $p_{\mathrm{T}} > 30$ GeV that do not come from the Higgs decay. Depending on the number of extra jets, three regions are defined: 0, 1 and at least 2 extra jets. Figure 1.6 shows the maximum split scheme which has 80 bins.

The *maximum splitting* categorization is only studied to provide a high flexibility in the choice of the bins to measure. The truth level categorization reflects the cuts at reconstructed level and covers the needs for possible BSM studies. The cross-section values of each bin in the maximum split categorization scheme are shown in Fig. 1.7. Due to the limited statistics, the contribution from most of the bins is too tiny to derive meaningful conclusions and a coarser scheme, denoted as *main splitting*, has been adopted. In the *resolved* analysis the categorization scheme includes five bins while the *boosted* analysis has only four bins. In both analyses, no split on the

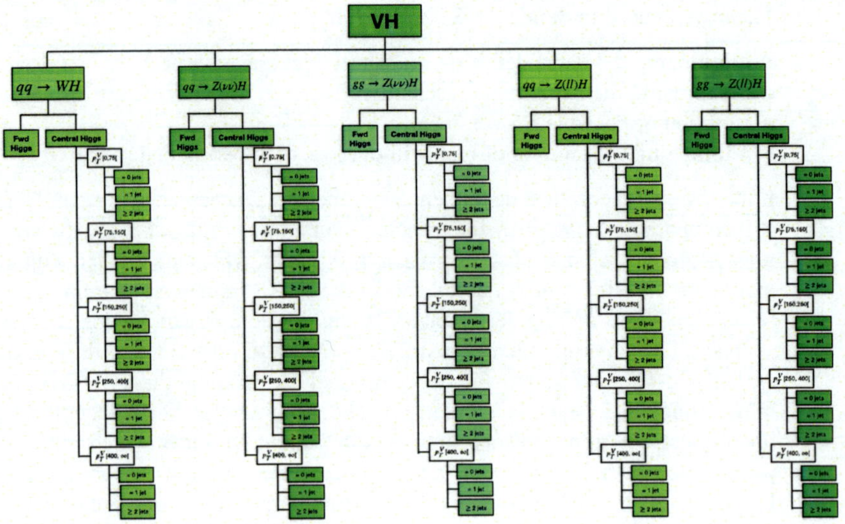

Fig. 1.6 VH STXS regions in the *maximum splitting* scheme. The boundaries of the truth p_{T}^{V} regions are expressed in GeV

number of extra jets is applied and all the ZH modes are merged. Only events in the central Higgs region ($|y_H| < 2.5$) are considered in the measurement. In the *resolved* analysis the splitting in p_{T}^{V} at 400 GeV is removed and the contributions of both ZH and WH for $p_{\mathrm{T}}^{V} < 75$ GeV are fixed to the SM together with the WH contribution for 75 GeV $< p_{\mathrm{T}}^{V} <$ 150 GeV. In the *boosted* analysis, the contributions of both production modes for $p_{\mathrm{T}}^{V} < 250$ GeV are fixed to the SM values. The *main splitting* categorizations for the *resolved* and *boosted* analysis are shown in Fig. 1.8.

1.5.1 STXS Measurements in the $VH(b\bar{b})$ Resolved Analysis

The VH simplified template cross-sections have been measured in the *resolved* analysis using the full Run 2 dataset [19]. The expected signal distributions in each STXS region are estimated from simulated signal samples by selecting events using the truth information, in particular the truth p_{T}^{V} called $p_{\mathrm{T}}^{V,\mathrm{t}}$. Figure 1.9 shows the expected signal yield and the expected signal fraction in each reconstructed event category for each STXS region. In 1-lepton and 2-lepton channels, almost all the events are from the WH and ZH production modes, respectively. In the 0-lepton there is a significant contribution ($\sim 15\%$) from the WH production modes. Most of the WH events reconstructed in the 0-lepton channel are events where the W boson decays in $\tau + \nu$, then the τ lepton decays hadronically and it is reconstructed as a jet.

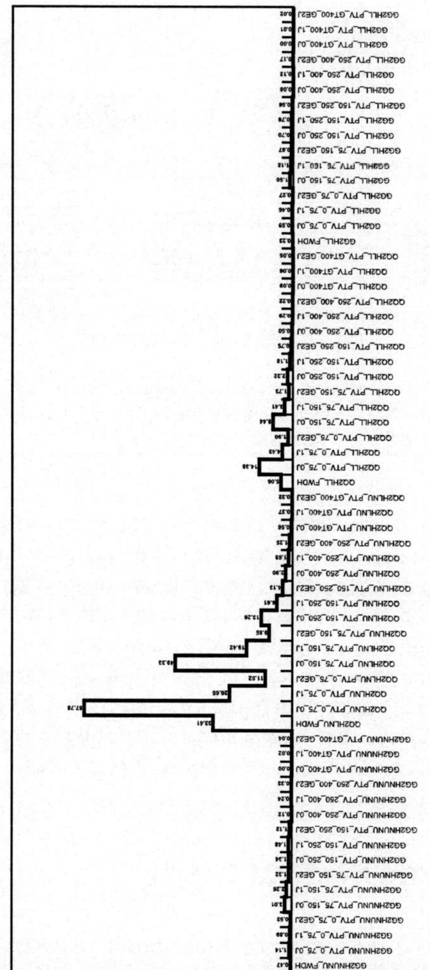

Fig. 1.7 Standard Model cross-sections in each region of the *maximum splitting* scheme at the centre-of-mass energy $\sqrt{s} = 13$ TeV

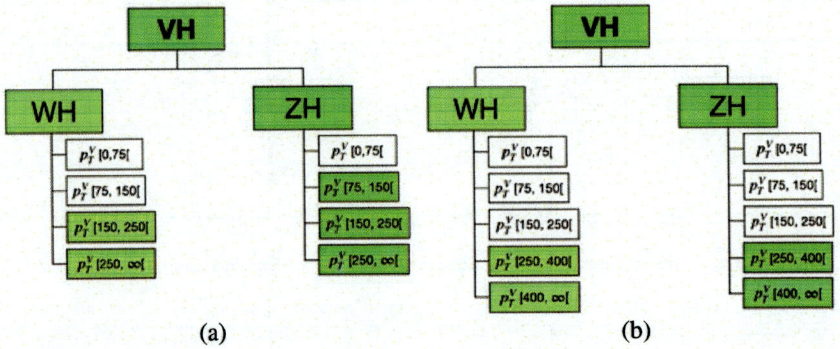

Fig. 1.8 VH STXS regions in the *main splitting* scheme used by the *resolved* (**a**) and *boosted* (**b**) analyses, respectively. The measurements are performed only in the bins shown in green. The bins shown in white are fixed to their SM values

The measured cross-section times branching fractions $\sigma(VH) \times BR(H \to b\bar{b}) \times BR(V \to \text{leptons})$, together with the SM predictions, in the five STXS regions under study are shown in Fig. 1.10. The results are in good agreement with the SM predictions and the cross-sections are measured with relative uncertainties varying from 30% in the highest p_{T}^{V} region to 85% in the lowest p_{T}^{V} regions. The largest uncertainties are statistical although in the lowest p_{T}^{V} regions systematics uncertainties make a sizeable contribution to the total uncertainty. For the ZH measurements, the signal uncertainties also make a sizeable contribution due to the limited precision of the theoretical calculations of the $gg \to ZH$ process.

1.6 The Effective Field Theory

Following the discovery of the Higgs boson, the Standard Model description has been tested looking for hints of BSM effects. Even if the scale Λ of the BSM physics is much larger than the typical SM scale, and larger than the energy scale at which the experiment is conducted, indication of new physics may rise by measuring deviations from the SM prediction. Effective Field Theories can be used to give a description of these effects. In this context the SM Lagrangian is extended into the SM Effective Field Theory (SMEFT) [21] considering a set of dimension D operators $\mathcal{O}_i^{(D)}$, larger than 4, where each consecutive term is suppressed by a larger power of Λ:

$$\mathcal{L}_{\mathrm{SMEFT}} = \mathcal{L}_{SM} + \sum_i \frac{c_i^{(5)}}{\Lambda}\mathcal{O}_i^{(5)} + \sum_i \frac{c_i^{(6)}}{\Lambda^2}\mathcal{O}_i^{(6)} + \sum_i \frac{c_i^{(7)}}{\Lambda^3}\mathcal{O}_i^{(7)} + \dots \tag{1.34}$$

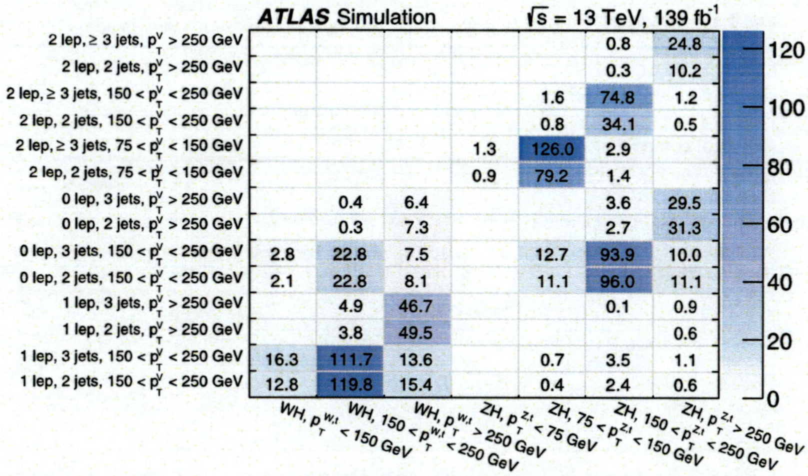

(a)

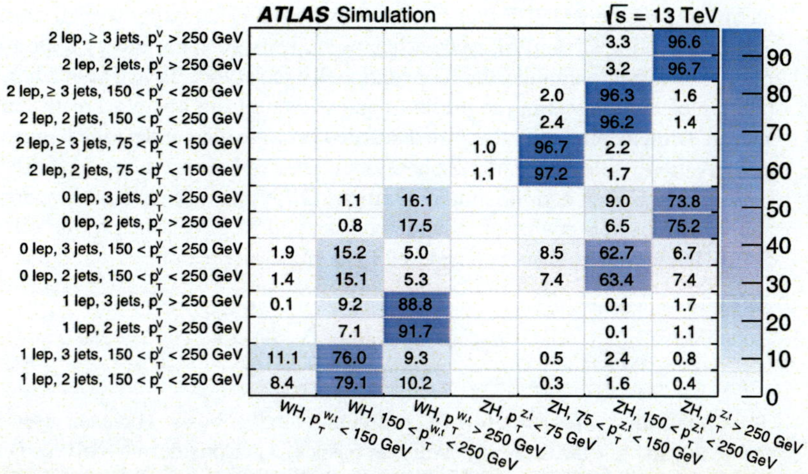

(b)

Fig. 1.9 Expected signal yields (**a**) and signal fractions (**b**) in each reconstructed category as a function of the STXS region. Events with event yield below 0.1 or signal fractions below 0.1% are not shown [19]

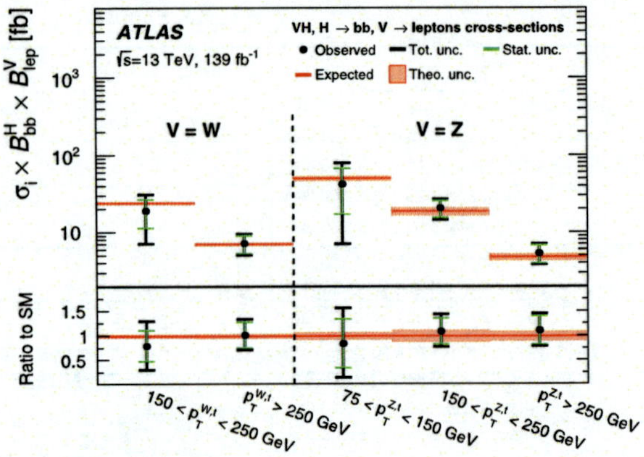

Fig. 1.10 Observed values of the $\sigma \times BR$ in the five STXS bins. The total observed uncertainty (black) is quoted together with the statistical component (green). The observation is compared to the SM prediction (red lines) and its uncertainty is represented by the red area [19]

where $\mathcal{L}_{SM}$ is the SM Lagrangian, Λ is the energy scale of the New Physics and c_i are the coupling constants called *Wilson coefficients*. The Wilson coefficients are free parameters of the SMEFT that can be constrained using experimental data. In the Standard Model, all the Wilson coefficients are zero. Also the energy scale Λ of new physics is a free parameter but conventionally is set to 1 TeV. The leading effects that are of the interest in the present context come from operators with dimension $D = 6$. Dimension-5 operators are discarded because they violate the lepton number conservation while dimension-7 operators violate the conservation of the difference between the barion and the lepton numbers. Differently, dimension-8 operators are suppressed by the power of Λ and are assumed to have a negligible effect. The SMEFT Lagrangian that is considered in the following can be re-written as:

$$\mathcal{L}_{\text{SMEFT}} = \mathcal{L}_{SM} + \sum_i \frac{c_i^{(6)}}{\Lambda^2} \mathcal{O}_i^{(6)} \tag{1.35}$$

The dimension-6 operators form a complete set called basis. The basis used to show the result in this thesis is the Warsaw basis [21]. Considering only dimension-6 operators, a total of 2499 additional operators can be built out of the SM field content. Only a small subset of these additional operators will affect the Higgs boson physics and 17 operators modify the $qq \to V(\to \text{leptons})H(\to b\bar{b})$ process, of which only 4 operators affect the $H \to b\bar{b}$ decay. The operators affecting the $VH(H \to b\bar{b})$ analysis are listed in Table 1.3 [21, 22] together with their Wilson coefficients. The ZH production cross-section is impacted by 13 operators, while the WH process depends only on 6 operators. In addition there are 3 operators that affect the total Higgs width Γ_H^{tot} and one that can induce possible variations of the $H \to b\bar{b}$ partial width. The operator to which the $VH(b\bar{b})$ *resolved* and *boosted* analyses are most sensitive is the one multiplied by the c_{Hq3} Wilson coefficient. Figure 1.11 shows examples of Feynman diagrams of this operator, showing which vertices are affected.

Table 1.3 Wilson coefficients and their corresponding dimension-6 operators in the Warsaw formulation affecting the $qq \to V(\to \text{leptons})H(\to b\bar{b})$ process [21, 22]

Wilson coefficient	Operator definition	Impact
$c_{H\Box}$	$(H^\dagger H)\Box(H^\dagger H)$	ZH, WH
c_{HDD}	$(H^\dagger D^\mu H)^*(H^\dagger D_\mu H)$	ZH, WH
c_{HW}	$H^\dagger H W^I_{\mu\nu} W^{I\mu\nu}$	ZH, WH
c_{HB}	$H^\dagger H B_{\mu\nu} B^{\mu\nu}$	ZH
c_{HWB}	$H^\dagger \tau^I H W^I_{\mu\nu} B^{\mu\nu}$	ZH
c_{Hl1}	$H^\dagger i\overleftrightarrow{D}_\mu H(\bar{l}_p\gamma^\mu l_r)$	ZH
c_{Hl3}	$(H^\dagger i\overleftrightarrow{D}^I_\mu H)(\bar{l}_p\tau^I\gamma^\mu l_r)$	ZH, WH
c_{He1}	$(H^\dagger i\overleftrightarrow{D}_\mu H)(\bar{e}_p\gamma^\mu e_r)$	ZH
c_{Hq1}	$(H^\dagger i\overleftrightarrow{D}_\mu H)(\bar{q}_p\gamma^\mu q_r)$	ZH
c_{Hq3}	$(H^\dagger i\overleftrightarrow{D}^I_\mu H)(\bar{q}_p\tau^I\gamma^\mu q_r)$	ZH, WH
c_{Hu}	$(H^\dagger i\overleftrightarrow{D}_\mu H)(\bar{u}_p\gamma^\mu u_r)$	ZH
c_{Hd}	$(H^\dagger i\overleftrightarrow{D}_\mu H)(\bar{d}_p\gamma^\mu d_r)$	ZH
c_{ll1}	$(\bar{l}_p\gamma_\mu l_r)(\bar{l}_s\gamma^\mu l_t)$	ZH, WH
$\lvert c_{dH}\rvert$	$(H^\dagger H)(\bar{q}_p d_r H)$	$\Gamma_H^{b\bar{b}}$
$\lvert c_{eH}\rvert$	$(H^\dagger H)(\bar{l}_p e_r H)$	Γ_H^{tot}
$\lvert c_{uH}\rvert$	$(H^\dagger H)(\bar{q}_p u_r \tilde{H})$	Γ_H^{tot}
c_{HG}	$H^\dagger H G^A_{\mu\nu} G^{A\mu\nu}$	Γ_H^{tot}

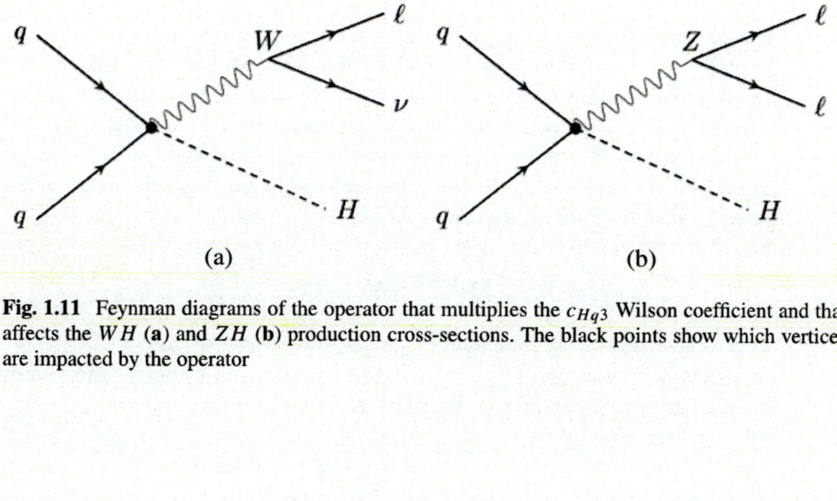

Fig. 1.11 Feynman diagrams of the operator that multiplies the c_{Hq3} Wilson coefficient and that affects the WH (**a**) and ZH (**b**) production cross-sections. The black points show which vertices are impacted by the operator

The ambitious physics program for the $VH(H \rightarrow b\bar{b})$ measurements is to determine the exclusion regions for these 17 parameters, with the best precision available.

Given the available statistics, one can find as a good compromise, the use of the STXS measurements to put limits on the EFT coefficients, profiting from the split in different kinematic STXS bins for the VH process.

In the SMEFT the expected cross-section values for different STXS bins become a function of the Wilson coefficients and they can be split into the following three contributions:

$$\sigma_{\text{SMEFT}} = \sigma_{\text{SM}} + \sigma_{\text{int}} + \sigma_{\text{BSM}} \tag{1.36}$$

where σ_{SM} is the cross-section value computed in the SM, σ_{int} arises through the interference of the SM and BSM processes and σ_{BSM} consists exclusively of BSM processes. The deviations from the SM predictions can then be expressed as a function of the Wilson coefficients:

$$\frac{\sigma_{\text{SMEFT}}}{\sigma_{\text{SM}}} = 1 + \sum_i A_i c_i + \sum_{i,j} B_{ij} c_i c_j \tag{1.37}$$

where A_i and B_{ij} are the coefficients defined as:

$$\begin{aligned} \frac{\sigma_{\text{int}}}{\sigma_{\text{SM}}} &= \sum_i A_i c_i \\ \frac{\sigma_{\text{BSM}}}{\sigma_{\text{SM}}} &= \sum_{ij} B_{ij} c_i c_j \end{aligned} \tag{1.38}$$

The coefficients A_i and B_{ij} are the linear and the quadratic terms in c_i, respectively, and they can be estimated for each STXS bin using Monte Carlo (MC) simulations. The interference terms are of the order of $1/\Lambda^2$, while the quadratic terms are of the order of $1/\Lambda^4$. Both linear and linear plus quadratic terms are studied.

The decay width of the Higgs boson in the SMEFT also suffers a dependency on the Wilson coefficients. In complete analogy to Eq. 1.36, the decay width into a given final state f can be decomposed as:

$$\Gamma^f_{\text{SMEFT}} = \Gamma^f_{\text{SM}} + \Gamma^f_{\text{int}} + \Gamma^f_{\text{BSM}} \tag{1.39}$$

where Γ^f_{SM} is the SM component, Γ^f_{int} is the interference component between the SM and BSM processes and Γ^f_{BSM} is the BSM component. As before the modifications to the decay width can be parametrised as:

$$\begin{aligned} \frac{\Gamma^f_{\text{int}}}{\Gamma^f_{\text{SM}}} &= \sum_i A^f_i c_i \\ \frac{\Gamma^f_{\text{BSM}}}{\Gamma^f_{\text{SM}}} &= \sum_{ij} B^f_{ij} c_{ij} \end{aligned} \tag{1.40}$$

The total Higgs boson width Γ is the sum of all the partial decay widths so it can be parametrised as:

$$\Gamma_{\text{SMEFT}} = \sum_f \Gamma^f_{\text{SMEFT}} = \sum_f \Gamma^f_{\text{SM}}(1 + \sum_i A^f_i c_i + \sum_{ij} B^f_{ij} c_i c_j) \tag{1.41}$$

Using the partial and total decay width definitions, the branching ratio can be written as a rational function of the Wilson coefficients.

Combining the parametrisation of the production cross-section $\sigma_{\text{SMEFT}}/\sigma_{\text{SM}}$, of the partial $\Gamma^f_{\text{SMEFT}}/\Gamma^f_{\text{SM}}$ and of the total $\Gamma_{\text{SMEFT}}/\Gamma_{\text{SM}}$ decay widths, the parametrisation of the production cross-section times branching ratio consists of three polynomials:

$$[\sigma(VH) \times BR(H \to b\bar{b})]_{\text{SMEFT}} = [\sigma(VH) \times BR(H \to b\bar{b})]_{\text{SM}} \cdot \frac{\sigma_{\text{SMEFT}}}{\sigma_{\text{SM}}} \cdot \frac{\Gamma^{b\bar{b}}_{\text{SMEFT}}/\Gamma^{b\bar{b}}_{\text{SM}}}{\Gamma_{\text{SMEFT}}/\Gamma_{\text{SM}}} \tag{1.42}$$

Maximum-likelihood fits are performed across the STXS regions to determine the Wilson coefficients. Limits on each individual Wilson coefficient are set by assuming all the others to vanish and one-dimensional confidence level (CL) intervals are inferred for the coefficient under study both with and without the quadratic term. Additionally, simultaneous fits with two parameters have been performed with and without the quadratic term. These fits allow to extract two-dimensional confidence levels for each pair of Wilson coefficients.

Attempting to simultaneously extract constraints on more than two coefficients leads to unmanageable correlations because there are less STXS regions than Wilson coefficients. An alternative approach, which can replace the one and the two-dimensional fit, is to use a linear combination of operators that are correlation-free to which the analysis is most sensitive. A well-defined simultaneous fit can be made to extract limits on the combinations of Wilson coefficients defined as eigenvectors. The combinations of Wilson coefficients are ordered in terms of experimental sensitivity. All the details to extract these linear combinations can be found in Ref. [22]. With this new approach it is possible to avoid very strong assumption that only a few operators contribute at time. The number of eigenvectors used in the analysis is equal to the number of STXS regions.

Many Wilson coefficients have degenerate effects on the VH STXS bins, especially the one from the Higgs decay. For this reason a linear parametrisation with explicit modifications of the branching ratio has been adopted:

$$[\sigma(VH) \times BR(H \to b\bar{b})]_{\text{SMEFT}} = [\sigma(VH) \times BR(H \to b\bar{b})]_{\text{SM}} \cdot \left[1 + \sum_i \alpha_i c_i + \mathcal{I}_{BR}\right] \tag{1.43}$$

Modifications to the branching ratio are not explicitly parametrised as functions of Wilson coefficient c_i, they are absorbed into the $\mathcal{I}_{BR}$ parameter which is introduced as an additional independent parameter. The $\mathcal{I}_{BR}$ parameter is defined as the ratio of the Higgs boson into a pair of b-quarks to its Standard Model predictions. In the next paragraph only the results of the one-dimensional and two-dimensional fits are shown. For completeness, the results of the eigenvector method are reported in Appendix A.

1.6.1 EFT Interpretation in the $VH(b\bar{b})$ Resolved Analysis

The largest sensitivity in the $VH(b\bar{b})$ *resolved* analysis is on the Wilson coefficients c_{HWB}, c_{HW}, c_{Hu}, c_{Hq3} [19]. In the following the limits on these coefficients are studied together with the limits on $|c_{dH}|$ coefficient which directly affects the $H \to b\bar{b}$ decay width. Figure 1.12 shows the ratio of the SMEFT cross-section times branching ratio over SM cross-section times branching ratio as a function of c_{Hu} in various STXS bins. The dashed lines include the linear parametrization only for the cross-section times branching ratio while the continuous lines use the linear + quadratic parametrization of these quantities. Since the c_{Hu} coefficient has an impact only on the ZH production, there is no variation of the cross-section times branching ratio with respect to the SM in the WH STXS bins. The plot shows that the linear approximation is fairly good for the STXS regions at low-p_T^V, while in the high-p_T^V STXS bin the quadratic terms gain importance with respect to the linear terms.

As already mentioned with only five STXS bins and 17 Wilson coefficients, a simultaneous fit to all the coefficients at the same time is not possible. To quantify how much the analysis phase space is sensitive to each operator, one-dimensional (1D) likelihood scans are performed considering only one operator at time. The full parametrisation of the cross-section times branching ratio is used in these 1D fits (Eq. 1.42). Figure 1.13 shows the observed (solid lines) and expected (dotted lines) profiles for the negative log-likelihood functions for the 1D fits of the c_{Hu} Wilson coefficient. The function that includes linear and quadratic parametrization shows the typical shape of the parabolic parametrisation with two minima.

The intervals at the 68% CL for the four Wilson coefficients to which the analysis is most sensitive and for $|c_{dH}|$, extracted from these likelihood scans, are shown

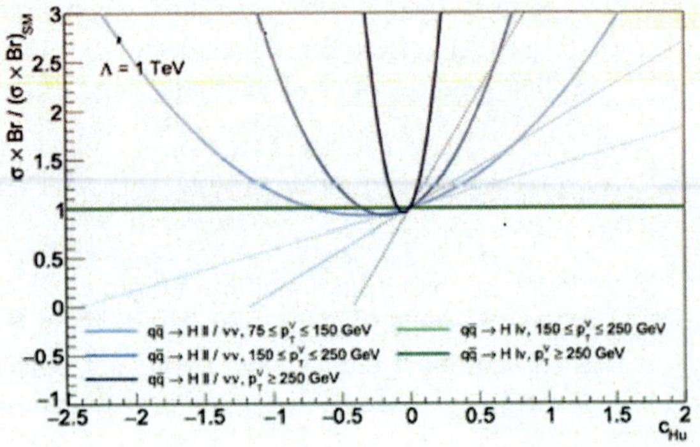

Fig. 1.12 Relative production cross-section times branching ratio into bottom quarks in various STXS bins as a function of the c_{Hu} Wilson coefficient. The solid curves use the linear + quadratic parametrisation for the VH production cross-section and Higgs decay widths while the dashed lines use only linear parametrisation

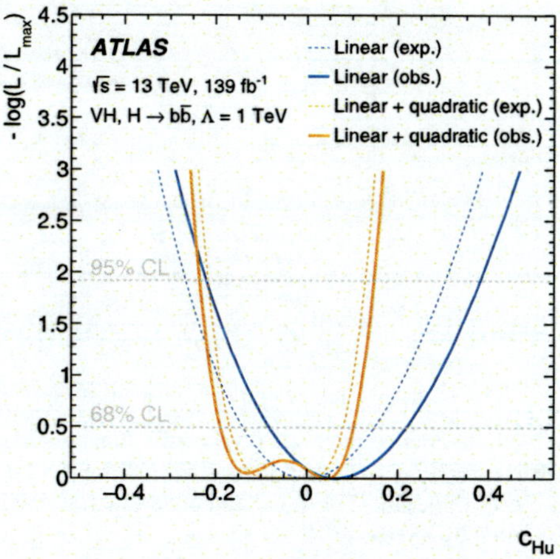

Fig. 1.13 Observed (solid lines) and expected (dotted lines) profiled negative log-likelihood functions for one-dimensional fits to constrain a single coefficient of an effective Lagrangian when the other coefficients are assumed to vanish. The coefficients c_{Hu} is shown for the case where only linear (blue lines) or linear and quadratic (orange lines) terms are considered [19]

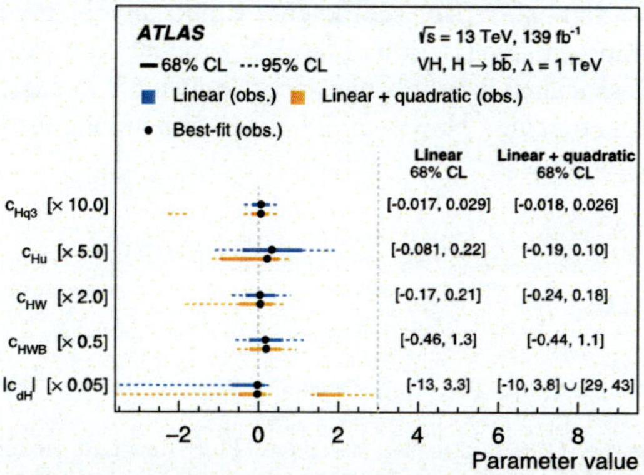

Fig. 1.14 Observed best-fit values and 1D confidence intervals for the Wilson coefficients to which this analysis has the greatest sensitivity and the $|c_{dH}|$ coefficient which directly affects the $H \to b\bar{b}$ decay width. Limits are shown for the case where only linear (blue) and linear and quadratic (orange) terms are considered. Confidence intervals are shown at both 68% CL (solid line) and 95% CL (dashed line) [19]

in Fig. 1.14 for the linear parametrisation and including the quadratic terms. The coefficient c_{Hq3} is constrained at 68% CL to be less than a few percent, while the constraints on the other three coefficients range from 10–30%. The constraint on the Wilson coefficient related to the branching ratio is much weaker. In most of the cases the constraints are found to depend on the presence of the quadratic terms.

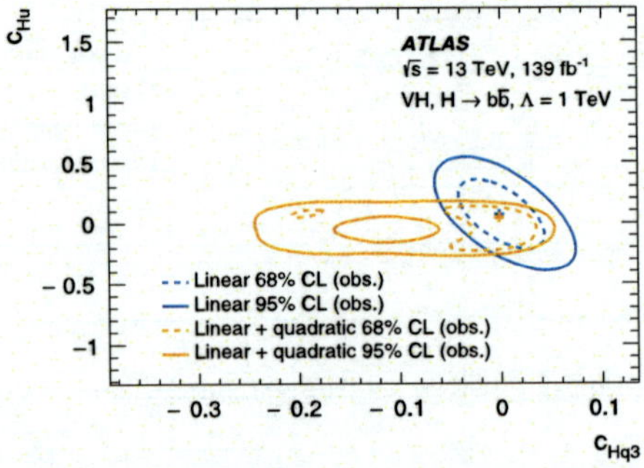

Fig. 1.15 Observed confidence intervals on the simultaneous likelihood to the pair of c_{Hq3} and c_{Hu} Wilson coefficients, at 68% (dashed lines) and 95% CL (solid lines). Limits are shown for the case where only linear (blue lines) or linear and quadratic (orange lines) terms are considered. The best points are marked by a cross [19]

Two-dimensional confidence intervals are derived for all the combinations of those four Wilson coefficients that the analysis can constrain best. Figure 1.15 shows the observed confidence interval at 68% (dashed lines) and 95% (solid lines) CL for the pair of c_{Hq3} and c_{Hu} Wilson coefficients. Limits are obtained from a linear and linear plus quadratic parametrisation of the VH production cross-section times branching ratio. The associated limits change appreciably upon inclusion of the quadratic terms in the parametrisation. However, the correlation among the two parameters is found very strong.

References

1. Glashow SL (1961) Partial symmetries of weak interactions. Nucl Phys 22:579. https://doi.org/10.1016/0029-5582(61)90469-2
2. Weinberg S (1967) A model of leptons. Phys Rev Lett 19:1264. https://doi.org/10.1103/PhysRevLett.19.1264
3. Salam A (1968) Weak and electromagnetic interactions. In: Svartholm N, Almqvist, Wiskell (eds) Proceedings of the 8th Nobel symposium, 1968, Conference proceedings C680519, vol 367. %7Bhttp://inspirehep.net/record/53083%7D
4. Englert F, Brout R (1964) Broken symmetry and the mass of Gauge vector mesons. Phys Rev Lett 13:321. https://doi.org/10.1103/PhysRevLett.13.321
5. Higgs PW (1964) Broken symmetries, massless particles and Gauge fields. Phys Lett 12:132. https://doi.org/10.1016/0031-9163(64)91136-9
6. Higgs PW (1964) Broken symmetries and the masses of Gauge Bosons. Phys Rev Lett 13:508. https://doi.org/10.1103/PhysRevLett.13.508
7. Dittmaier S et al. (2011) Handbook of LHC Higgs cross sections: 1. Inclusive observables (2011). https://doi.org/10.5170/CERN-2011-002. arXiv:1101.0593 [hep-ph]
8. Tanabashi MEA (2018) Review of particle physics. Phys Rev D 98:030001. https://doi.org/10.1103/PhysRevD.98.030001

9. ATLAS Collaboration (2012) Observation of a new particle in the search for the Standard Model Higgs boson with the ATLAS detector at the LHC. Phys Lett B 716:1. https://doi.org/10.1016/j.physletb.2012.08.020. arXiv:1207.7214 [hep-ex]
10. CMS Collaboration (2012) Observation of a new boson at a mass of 125 GeV with the CMS experiment at the LHC. Phys Lett B 716:30. https://doi.org/10.1016/j.physletb.2012.08.021. arXiv:1207.7235 [hep-ex]
11. Biekötter A, Knochel A, Krämer M, Liu D, Riva F (2015) Vices and virtues of Higgs effective field theories at large energy. Phys Rev D 91:055029. https://doi.org/10.1103/PhysRevD.91.055029. arXiv:1406.7320 [hep-ph]
12. ATLAS Collaboration (2017) Evidence for the $H \to b\bar{b}$ decay with the ATLAS detector. JHEP 12:024. https://doi.org/10.1007/JHEP12(2017)024. arXiv:1708.03299 [hep-ex]
13. Collaboration CMS (2018) Evidence for the Higgs boson decay to a bottom quark-antiquark pair. Phys Lett B 780:501. https://doi.org/10.1016/j.physletb.2018.02.050. arXiv:1709.07497 [hep-ex]
14. ATLAS Collaboration (2018) Observation of $H \to b\bar{b}$ decays and VH production with the ATLAS detector. Phys Lett B 786:59. https://doi.org/10.1016/j.physletb.2018.09.013. arXiv:1808.08238 [hep-ex]
15. Collaboration CMS (2018) Observation of Higgs Boson decay to bottom quarks. Phys Rev Lett 121:121801. https://doi.org/10.1103/PhysRevLett.121.121801. arXiv:1808.08242 [hep-ex]
16. de Florian D et al. (2016) Handbook of LHC Higgs cross sections: 4. Deciphering the nature of the Higgs sector, 2/2017 (2016). https://doi.org/10.23731/CYRM-2017-002. arXiv:1610.07922 [hep-ph]
17. Andersen J et al. (2016) Les Houches 2015: physics at TeV Colliders Standard Model working Group Report. In: 9th Les Houches workshop on physics at TeV colliders. arXiv:1605.04692 [hep-ph]
18. ATLAS Collaboration (2019) Measurement of VH, $H \to b\bar{b}$ production as a function of the vector-boson transverse momentum in 13 TeV pp collisions with the ATLAS detector. JHEP 05:141. https://doi.org/10.1007/JHEP05(2019)141. arXiv:1903.04618 [hep-ex]
19. Aad G et al. (2020) Measurements of WH and ZH production in the $H \to b\bar{b}$ decay channel in pp collisions at 13 TeV with the ATLAS detector (2020). arXiv:2007.02873 [hep-ex]
20. Aad G et al. (2020) Measurement of the associated production of a Higgs boson decaying into b-quarks with a vector boson at high transverse momentum in pp collisions at $\sqrt{s} = 13$ TeV with the ATLAS detector. arXiv:2008.02508 [hep-ex]
21. Grzadkowski B, Iskrzynski M, Misiak M, Rosiek J (2010) Dimension-six terms in the standard model Lagrangian. JHEP 10:085. https://doi.org/10.1007/JHEP10(2010)085. arXiv:1008.4884 [hep-ph]
22. Methodology for EFT interpretation of Higgs boson Simplified Template Cross-section results in ATLAS, Technical report. ATL-PHYS-PUB-2019-042, CERN (2019). https://cds.cern.ch/record/2694284

Chapter 2
The Large Hadron Collider and the ATLAS Experiment

This chapter is dedicated to the two key ingredients necessary for data analysis in high energy physics: the particle accelerator and the detector. The first part of the chapter is fully dedicated to the world's largest particle accelerator, the Large Hadron Collider (LHC), while the second part is focused on the ATLAS detector with a brief description of all the sub-detector components.

2.1 The LHC

The Large Hadron Collider (LHC) [1] is a superconducting-hadron accelerator and collider installed in the existing 27 km tunnel for the Large Electron Positron (LEP) machine [2] on the French-Swiss border at CERN laboratory. It is a circular accelerator for protons and heavy ions that makes collisions with an energy at the center-of-mass $\sqrt{s}$ up to 13 TeV.

The collider tunnel contains two adjacent parallel beamlines called beam pipes each containing a beam. The beams travel in opposite directions around the ring before they collide. The LHC relies on 1232 superconducting niobium-titanium dipole magnets at the edge of the present technology to bend the path of the proton beams travelling through the ring. The dipole magnets are cooled to a temperature down to 1.9 K using superfluid helium, and they provides a magnetic field of 8 T generated by an electric current at 11700 A. To stabilize and focus the beam, additional 392 quadrupole magnets complement the dipole system.

The process of acceleration is obtained with a chain of machines that accelerate particles to higher energy step-by-step as shown in Fig. 2.1. The proton source is hydrogen gas. An electric field is used to strip hydrogen atoms of their electrons to yield protons. The protons are then injected into the first accelerator Linac 2, a linear accelerator, that accelerates protons up to 50 MeV. The beam is then injected into the Booster, a circular accelerator which accelerates particles to 1.4 GeV. Once protons reach this energy they are injected into the Proton Synchroton (PS), which

G. Di Gregorio, *A Beauty-ful Boson*, Springer Theses,
https://doi.org/10.1007/978-3-031-20013-7_2

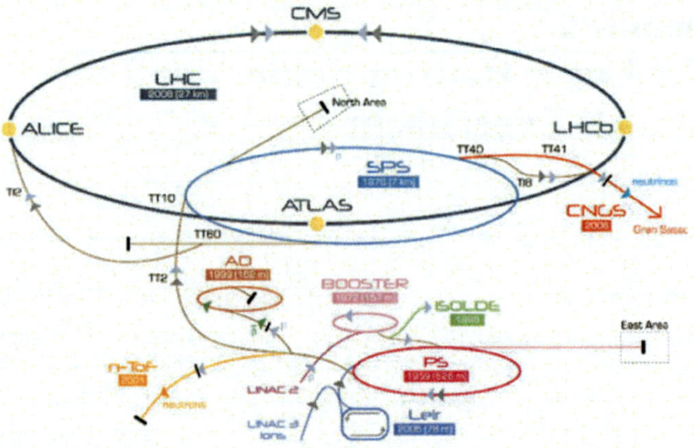

Fig. 2.1 Schematic view of the accelerator system at CERN with the four collision points and other beams extracted at different accelerator level

pushes the beam to 25 GeV. Protons are then sent into the Super Proton Synchroton (SPS) where they are accelerated to 450 GeV. The protons are finally transferred to two beam pipes of LHC. Two proton beams circulate in opposite directions and they are accelerated up to the energy of 6.5 TeV. The two beams collide at four interaction points where ATLAS, CMS, LHCb and ALICE detectors are installed. The centre-of-mass energy at the collision point is $\sqrt{s}$ =13 TeV. This thesis uses the data collected by the ATLAS experiment which is discussed in more details in Sect. 2.2.

The accelerator complex also includes the Antiproton Decelerator (AD) and the Online Isotope Mass Separator (ISOLDE) facility, a facility that generates neutrinos for the CERN Neutrinos to Gran Sasso (CNGS) project, as well as the neutron time-of-flight facility (nTOF).

As already mentioned, protons are not the only particles accelerated in the LHC, also heavy ions are accelerated. The heavy ions are produced from a source of vaporised lead and enter Linac 3 before being collected and accelerated in the Low Energy Ion Ring (LEIR). Then they follow the same route to the maximum energy as the protons. Figure 2.1 shows the accelerator chain, the collisions points with the four LHC experiments and the other beams extracted at different accelerator levels.

The LHC was designed to collide proton beams with a center-of-mass energy up to 14 TeV. During the first LHC run, the so-called Run 1, lasted from 2011 to 2012, LHC delivered proton-proton (*p*-*p*) collisions at center-of-mass energy of 7 TeV and 8 TeV. In 2013 and 2014, there was the shutdown for consolidation activities. After the shutdown, there was the second run, the so-called Run 2, started in 2015 and lasted until the end of 2018. During the Run 2, the center-of-mass energy of the *p*-*p* collisions was 13 TeV.

The protons circulating in the LHC ring are arranged in *bunches*, and each beam is formed by a train of up to 2808 bunches with approximately 10^{11} protons per bunch.

The separation between bunches is called *bunch spacing*. The bunch spacing was 50 ns during the Run 1 and 25 ns during the Run 2.

Luminosity

The number of events for a given process generated in LHC collisions is given by the formula:

$$N_{event} = \sigma_{event}\,\mathcal{L} \tag{2.1}$$

where σ_{event} is the cross-section for the process under study and $\mathcal{L}$ is the integrated luminosity. The former factor depends on the considered physics process and it is proportional to the probability of a specific final state to appear. More formally, the cross-section for a process is defined as the number of interactions per unit time per target particle divided by the incident flux [3]. It is usually measured in barn (1 b = 10^{-24} cm^2). This quantity may depend on energy in different ways for different processes: when energy increases, new phenomena contributing to the cross-section may appear and become dominant and, at the same time, others can be suppressed. The integrated luminosity is a parameter of the machine and it is defined as the time integral of the instantaneous luminosity L. The instantaneous luminosity can be expressed as a function of the geometrical characteristics of the colliding bunches and of the machine parameters:

$$L = \frac{N_B^2 n_b f_{rev} \gamma_r}{4\pi \epsilon_n \beta^*} F \tag{2.2}$$

where N_b is the number of protons per bunch, n_b is the number of bunches per beam, f_{rev} is the revolution frequency, γ_r is the relativistic gamma factor, ϵ_n is the normalised transverse beam emittance,[1] β^* is the beta function[2] at the collision point and F is the geometrical luminosity reduction factor due to the crossing angle at the interaction point. The geometrical luminosity reduction factor F is defined as:

$$F = \left(1 + \left(\frac{\theta_c \sigma_z}{2\sigma^*}\right)^2\right)^{-\frac{1}{2}} \tag{2.3}$$

where θ_c is the full crossing angle at the interaction point, σ_z is the RMS of the bunch length and σ^* is the RMS of the transverse beam size at the interaction point. The values of the main accelerator parameters are shown in Table 2.1. The instantaneous luminosity is measured in pb s^{-1} or cm^{-2} s^{-1} units. The LHC target instantaneous luminosity is $L = 10^{34}$ cm^{-2} s^{-1}.

[1] The normalised transverse beam emittance represents the average spread of the beam in the momentum-position phase space transverse to the beam.

[2] The beta function describes the size of the beam in the plane transverse to the beam. β^* refers to this size at the interaction point.

Table 2.1 Summary of the nominal design LHC parameters

Variable	Value	Description
N_b	10^{10}–10^{11}	Number of particles per bunch
n_b	2808	Number of bunches per beam
f_{rev}	11245 Hz	Revolution frequency
γ_r	~ 7000	Relativistic gamma factor
ϵ_n	3.75 μm	Normalized transverse beam emittance
β^*	0.55 m	Beta function at the collision point
F	0.84	Geometric luminosity reduction factor
θ_c	±285 μrad	Crossing angle at interaction point
σ_z	7.55 cm	RMS bunch length
σ^*	16.7 μm	RMS beam size

Equation 2.1 shows a linear dependency between the number of events and the integrated luminosity. Higher luminosity means larger amount of primary interactions: this is very important because it allows the study of rare events.

During the Run 2 LHC delivered to ATLAS 156 fb of integrated luminosity, 146 fb were recorded by the experiment and 139 fb were used for physics analyses after certification of good quality data (Fig. 2.2a).

Due to the high instantaneous luminosity at LHC, a large number of proton interactions per bunch crossing are produced. During Run 2 period, the mean number of interactions per bunch crossing $< \mu >$ is varied between 34 and 38 (Fig. 2.2b). These additional interactions per bunch crossing are called *in-time pile-up*. Due to the high granularity of the detectors, it is possible to isolate the interaction of interest from the in-time pile-up by distinguishing among different interaction vertices and their associated particles. If the bunch crossing frequency is high, it may happen that interactions from adjacent bunch crossing will fall inside the detector integration time and they will be recorded in the same event. This phenomenon is indicated as *out-of-time pile-up*.

LHC Experiments

In the LHC ring the two beams collide in four points which correspond to the four detectors designed for different physics programs. These detectors are:

- **A Large Ion Collider Experiment (ALICE)** [5] is a detector which is designed to study the physics of strongly interacting matter at extreme energy densities, generated in high energy heavy ion collisions. Lead ions collisions recreate in the laboratory conditions similar to those just after the Big Bang, where protons and neutrons "melt", freeing the quarks from their bonds with gluons. This is the quark-gluon plasma. The existence of such a phase and its properties are key issues in the the theory of quantum chromodynamics. The ALICE Collaboration looks for the quark-gluon plasma observing how it progressively gives rise to the particles that constitute our universe today.

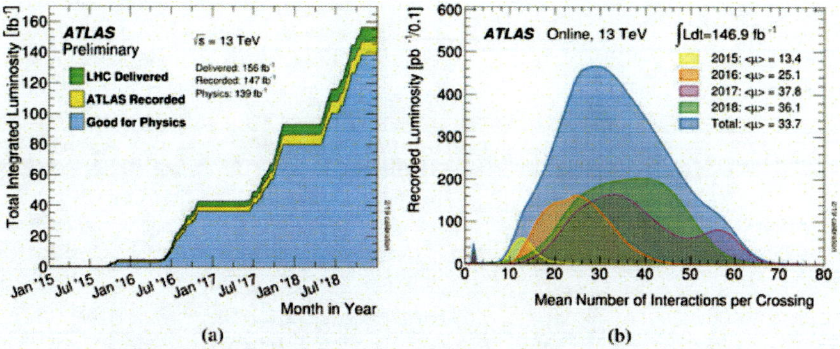

Fig. 2.2 **a** Total integrated luminosity versus time delivered to ATLAS (green), recorded by ATLAS (yellow), and certified to be good quality data (blue) during stable beam proton-proton collisions at 13 TeV center-of-mass energy in 2015–2018. **b** Mean number of interactions per bunch crossing for the fill Run 2 period [4]

- **A Toroidal LHC Apparatus (ATLAS)** [6] is one of the two general-purpose detector at LHC. It investigates a wide range of physics, from measurements of the Higgs boson properties to the search of the Higgs boson to extra dimensions and particles that could make up dark matter. Its name is given by the toroidal magnet placed in the outer part of the detector.
- **Compact Muon Solenoid (CMS)** [7] is the other LHC general-purpose detector with a scientific program very similar to ATLAS. Even if it has the same scientific goal of the other multi-purpose detector, it uses different technical solution and a different magnet-system design: the CMS detector is built around a huge solenoid magnet which provides an axial magnetic field of 3.8 T both for the tracker and the muon system.
- **Large Hadron Collider beauty (LHCb)** [8] is a detector focused on flavour physics, performing precise measurements of CP violations and rare decays of bottom and charm hadrons. The LHCb experiment is asymmetric with respect to the interaction point and it uses a series of sub-detectors to detect mainly forward particles. The first sub-detector is mounted close to the collision point and the others are placed in sequence in a line aside the beam orbit over a total length of 20 m.

2.2 The ATLAS Experiment

ATLAS (A Toroidal LHC ApparatuS) [6] is one of the four main experiments taking data at LHC and it is an example of a general purpose detector. This experiment is housed in a hall about 100 m underground, in a beam interaction point of the LHC ring. The detector has a cylindrical shape around the beam pipe with a diameter

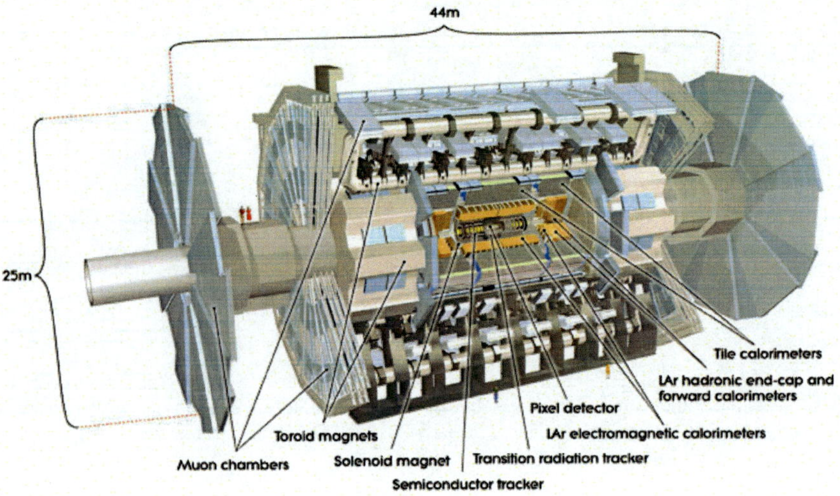

Fig. 2.3 Cut-away view of the ATLAS detector [6]

of 25 m and a length of 44 m. The ATLAS detector consists of a series of sub-detectors placed around the beam pipe in layers with cylindrical symmetry. Near the beam pipe there is the innermost sub-system, the Inner Detector (ID) which is surrounded by a solenoidal magnet. Then there is the calorimeter system, formed by the electromagnetic (EM) and the hadronic calorimeters. The outer sub-detector is the Muon Spectrometer (MS). Three toroids are situated outside the calorimeters and within the MS. Additionally, two end-caps, perpendicular to the beam, are situated at the end of the detector to improve the detector coverage in the forward region. A schematic view of the ATLAS detector with all the sub-detectors is shown in Fig. 2.3.

Coordinate System

The coordinate system and nomenclature used to describe the ATLAS detector and the particles emerging from the *p*-*p* collisions are briefly summarized here. The interaction point is defined as the origin of the coordinate system, the beam direction coincides with the z-axis and the x-y plane is the transverse plane to the beam direction. A cylindrical coordinate system is used to describe the detector: the azimuthal angle ϕ is measured around the beam axis and the polar angle θ is the angle from the beam axis. Usually, the θ coordinate is replaced by the pseudorapidity variable defined as

$$\eta = -\ln \tan\left(\frac{\theta}{2}\right) \tag{2.4}$$

In the massless particle limit, the pseudorapidity is equivalent to the rapidity $y = \frac{1}{2}\log\left(\frac{E+p_z}{E-p_z}\right)$. A particle with $\eta = 0$ is travelling perpendicular to the beam pipe

($\theta = 90°$), while if the particle is travelling parallel to the beam pipe ($\theta = 0°$ or $\theta = 180°$), its pseudorapidity tends to infinity ($|\eta| \rightarrow \infty$).

Another quantity used in this thesis is the angular distance between two objects i and j which is defined as:

$$\Delta R = \sqrt{(\eta_i - \eta_j)^2 + (\phi_i - \phi_j)^2} \tag{2.5}$$

where η_i and η_j are the pseudorapidity of the two objects and, ϕ_i and ϕ_j are their azimuthal angles.

The momentum of an object is expressed in the Cartesian coordinates as $\boldsymbol{p} = (p_x, p_y, p_z)$, where p_x, p_y, p_z are the momentum in the x, y and z directions. Often properties are given in the transverse plane to the beam (x-y plane). For example the transverse momentum p_{T} of a particle is defined as:

$$p_{\mathrm{T}} = \sqrt{p_x^2 + p_y^2} = |\mathbf{p}| \sin\theta \tag{2.6}$$

2.2.1 Magnet System

One of the key feature of the ATLAS detector is the hybrid superconducting magnet system which consists of a solenoid, a barrel toroid and two end-cap toroids, as shown in Fig. 2.4. The purpose of the magnet system is to provide the magnetic field for the ID and the MS in order to curve the trajectory of charge particles.

The solenoid is aligned on the beam axis and it is positioned between the ID and the EM calorimeter. The inner and the outer diameters of the solenoid are 2.46 m and 2.56 m and its axial length is 5.8 m. It provides a 2 T axial magnetic field for the central tracker.

The toroid provides a magnetic field for the MS and it is formed by eight coils located outside the calorimeters and within the MS. The magnetic field goes through 25.3 m in length, with inner and outer diameter of 9.4 m and 20.1 m, respectively. Also the two end-caps toroids are formed by eight coils and they are situated at the end of the barrel toroid; they have inner and outer diameter of 1.65 m and 10.7 m

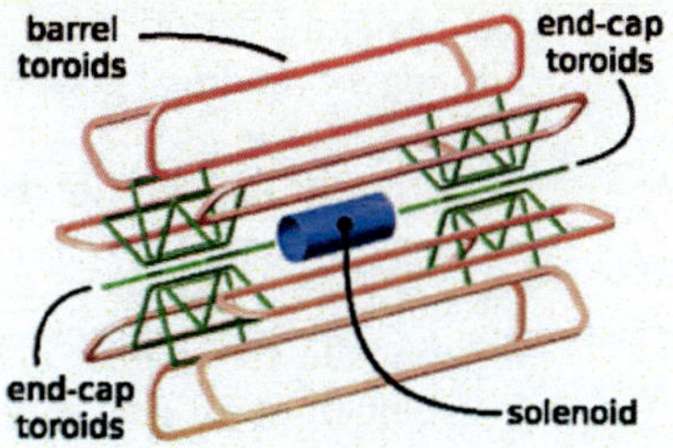

Fig. 2.4 Geometry of ATLAS magnet system: the solenoid (blue), the barrel toroid (pink) and the two end-cap toroids (green) [9]

and a length of 5 m. The barrel toroid provides a magnetic bending over the range $|\eta| < 1.4$, while the magnetic bending of the end-cap toroids is over the range $1.6 < |\eta| < 2.7$. The region over $1.4 < |\eta| < 1.6$ is usually called *transition region*. Here the magnetic deflection is given by a combination of barrel and end-cap fields. The three superconducting toroids generate a magnetic field for the muon detectors. In particular this configuration creates a magnetic field mostly orthogonal to the muon trajectory in order to minimize the degradation of resolution. The toroidal magnetic field is not uniform: the barrel toroid provides 2–6 Tm while the end-cap toroids contributes with 4–8 Tm. The transition region is characterized by lower bending power.

2.2.2 Inner Detector

The ATLAS Inner Detector is designed to identify charged particles, to measure their momentum and to provide both primary and secondary vertex measurements for tracks with the transverse momentum p_T above 0.5 GeV. The ID operates in a 2T magnetic field generated by the central solenoid which bends the trajectory of charged particles. The region covered by the ID is $|\eta| < 2.5$.

Since it is the nearest detector from the collision point, the ID is designed to resolve a large track density. To achieve the required momentum and vertex resolutions, the ID is made by high granularity sub-detectors. The ID consists of four tracking devices: the Inserable B-Layer (IBL), the Pixel detector, the Semi-Conductor Tracker (SCT) and the Transition Radiation Tracker (TRT). The ID detector is divided in three separate regions: the barrel section and the two end-caps sections. In the barrel section the four sub-modules are arranged on concentric cylinders around the beam axis, while in the end-cap regions there are only the Pixel detector, the SCT and the TRT which are located on disks perpendicular to the beam axis. The layout of the ID is illustrated in Fig. 2.5.

The IBL [11] was installed during the long shut-down of LHC between the Run 1 and Run 2 period and it is the inner-most layer of the ATLAS detector. The main motivation of the IBL is to improve the robustness of the tracking due to dead pixel modules, the tracking precision and the resilience of tracking performance for the increasing luminosity. For example, the addition of the IBL improved the resolution of the track impact parameters by about 40% for tracks with a $p_T < 1$ GeV, bettering the *b*-tagging performance. The IBL consists of a unique layer of pixel modules forming a cylindrical detector of radius 3.3 cm around the interaction and offering a coverage of $|\eta| < 2.5$.

The next sub-detector is the Pixel detector which consists of pixel modules arranged in three concentric cylinders in the barrel region covering $|\eta| < 1.5$. In the end-cap regions the Pixel detector is composed of three disks of silicon modules covering the $1.5 < |\eta| < 2.5$ region. Pixels are 50×400 μm^2 in $R - \phi \times z$, giving a resolution of 10 μm in the transverse direction and 115 μm in the longitudinal direction.

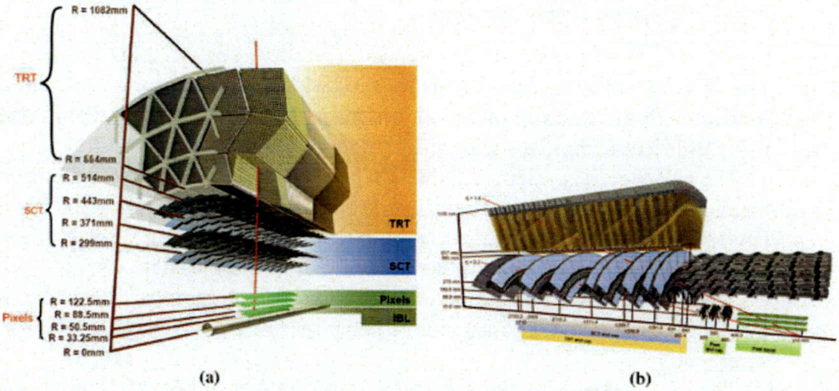

Fig. 2.5 **a** Cut-away of the ATLAS ID barrel diagram made up by the IBL, the three pixel layers, the four layers of the SCT and the straw layers of the TRT [10]. **b** Cut-away of the ID end-cap made up by three pixel disks, nine disks of the SCT and TRT wheels [6]

The next layer is the SCT which consists of four layers of microstrip pairs, covering 255 mm $< R <$ 549 mm in the barrel region ($|\eta| < 1.4$.). In the end-cap region ($1.4 < |\eta| < 2.5$), the SCT consists of two sets of 9 disk layers, covering 275 mm $< R <$ 560 mm for 839 mm $< |z| <$ 2735 mm and giving a resolution of 17 μm in the transverse direction and 580 μm in the longitudinal direction. In this region silicon strip were chosen over silicon pixels due to the lower expected particle density. This allows to maintain a good spatial resolution while containing the number of readout channels.

The last layer of the ID is the TRT which consists of about 300,000 drift tubes (straws) filled by xenon gas with a diameter of 4 mm. When a charged particle crosses the TRT straws, it ionises the active gas and produces primary ionisation clusters in the gas. The straw wall at high negative voltage creates an electric field which accelerates electrons towards the central anode and generates more electrons producing a detectable current signal. In the barrel the straws are placed parallel to the beam pipe and they cover the 560 mm $< R <$ 1080 mm and $|z| <$ 720 mm region. In the two end-caps the straws are arranged perpendicular to the beam pipe and they cover 827 mm $< |z| <$ 2774 mm and 617 mm $< R <$ 1106 mm. The TRT covers $|\eta| < 2.0$, providing $R - \phi$ information, with an accuracy of 130 μm per straw and $\sim$ 35 hits per track.

Measuring the curvature path and the direction of the charged particle from the hits in the ID, it is possible to determine the momentum and the charge of the particle. The target track momentum resolution for a particle with transverse momentum p_{T} is given by:

$$\frac{\sigma_{p_{\mathrm{T}}}}{p_{\mathrm{T}}} = 0.05\% p_{\mathrm{T}} \oplus 1\% \tag{2.7}$$

2.2.3 Calorimetric System

The ATLAS calorimetric system has the task to measure energy of particles that interact with the material of which it is made. It covers a huge η region ($|\eta| < 4.9$) using different techniques with different resolutions.

The calorimetric system is divided in two sections: the electromagnetic calorimeter and the hadronic calorimeter. The first one provides a measurement of electromagnetic showers, while the second one provides a measurement of hadronic showers. Figure 2.6 shows the layout of the ATLAS calorimetric system.

The ATLAS calorimeters consist of a number of sampling detectors with full ϕ-symmetry and coverage around the beam axis in order to have a good $E_{\mathrm{T}}^{\mathrm{miss}}$ measurement.

The EM calorimeter is a sampling calorimeter made of steel cladded lead absorbers, arranged in an accordion structure and plunged in the liquid argon (LAr). It is divided in three regions: the barrel part covers the range $|\eta| < 1.47$ and the two end-cap parts cover the range $1.375 < |\eta| < 3.2$. Interacting particles with the EM calorimeter ionize the liquid argon and this allows to measure their energy. The evolution of an electromagnetic shower is characterized by the material's *radiation length* X_0 which is defined as the mean distance over which the particle loses $\frac{1}{e}$ of its energy. The total thickness of the EM calorimeter is more than 22 X_0 in the barrel part and more than 24 X_0 in the end-caps.

In the region $|\eta| < 1.8$, a pre-sampler calorimeter is used to estimate for the energy lost by electrons and photons upstream of the calorimeter. This pre-sampler consists of an active LAr layer 1.1 cm thick in the barrel region and 0.5 cm thick in the end-cap region.

The hadronic calorimeter is also a sampling calorimeter. It is composed by three parts: the Tile Calorimeter (TileCal), the Hadronic End-cap Calorimeter (HEC) and

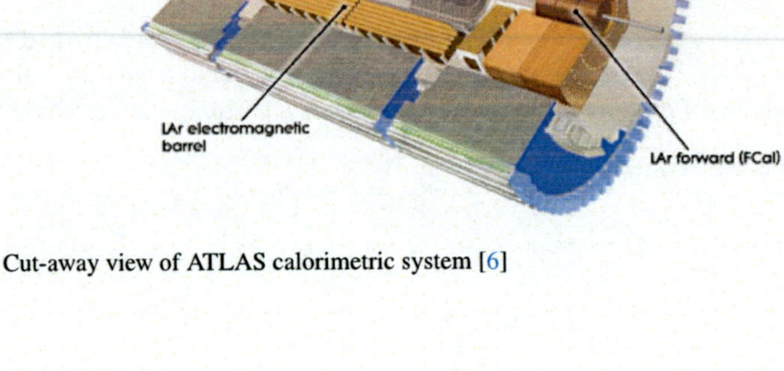

Fig. 2.6 Cut-away view of ATLAS calorimetric system [6]

Table 2.2 General performance goals of the ATLAS sub-detectors with their coverage [6]

Sub-detector	Resolution
EM calorimeter	$\sigma_E/E = 10\% /\sqrt{E} \oplus 0.7\%$
TileCal, HEC	$\sigma_E/E = 50\%/\sqrt{E} \oplus 3\%$
FCal	$\sigma_E/E = 100\%/\sqrt{E} \oplus 10\%$

the Forward Calorimeter (FCal). The shape of the hadronic shower is different from the shape of the electromagnetic shower due to the nature of the strong interaction with respect to the electromagnetic interaction. The characteristics depth of a material is the *interaction length* λ which represents the mean distance travelled by a hadron until its energy has been reduced by a fraction $\frac{1}{e}$ of the initial energy. The total thickness of the hadronic calorimeter is more than 11 λ.

The TileCal is placed around the EM calorimeter envelopes and it is made by steel plates as absorbers, and plastic scintillating tiles as active materials. It is composed by three barrels: the central barrel covers the region $|\eta| < 1.0$ and the two extended barrels cover the range $0.8 < |\eta| < 1.7$. A complete description of TileCal can be found in Appendix B, together with the stability studies of the photomultipliers reading out TileCal.

The HEC consists of two independent wheels per end-cap, located behind the end-cap electromagnetic calorimeter. They cover the region $1.5 < |\eta| < 3.2$ and they use the liquid argon technology with copper plates.

The FCal is another sampling calorimeter made by copper and tungsten as absorbers and liquid argon as active material. It is integrated into the end-cap cryostats and it covers the region $3.1 < |\eta| < 5$.

Even if the calorimetric structure consists of many detectors, it offers high hermeticity and granularity. Excellent energy resolution is required in many analysis. The target ATLAS calorimeter resolutions are summarized in Table 2.2.

2.2.4 Muon Spectrometer

The outermost and largest system of the ATLAS detector is the Muon Spectrometer. Due to the larger mass of muons over electrons, the energy lost for Bremsstrahlung is reduced and the muons pass through the ATLAS calorimeters with minimal interactions. The tracking chambers are used to measure the path of muons which is deviated by the magnetic field. The layout of the MS is shown in Fig. 2.7.

This system measures the muon momentum in a pseudorapidity range $|\eta| < 2.7$ thanks to the magnetic bending. The magnetic configuration creates a field which is orthogonal to the muon trajectories while minimizing the degradation of resolution due to multiple scattering. The muon system is also designed to trigger on muons in the region $|\eta| < 1.4$.

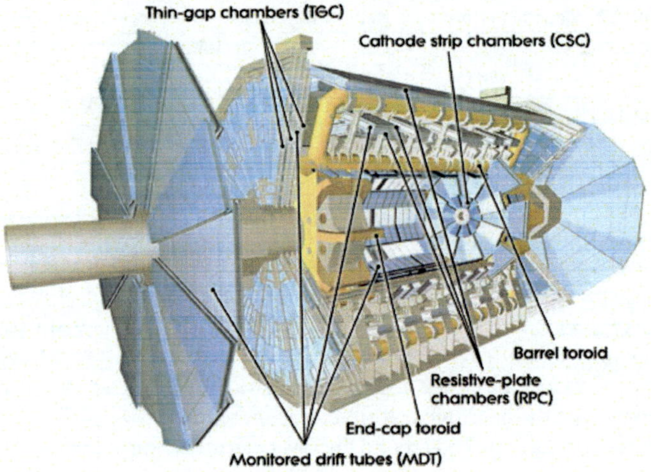

Fig. 2.7 Cut-away view of ATLAS muon system

In the barrel region ($|\eta| < 1.4$) tracks are measured in chambers arranged in three cylindrical layers around the beam axis at radii of approximately 5, 7.5 and 10 m; in the transition and in the end-cap regions ($|\eta| > 1.6$) the chambers are installed in planes perpendicular to the beam and located at distances of $|z| \approx 7.4$, 10.8, 14 and 21.5 m from the interaction point.

Four types of tracking chambers with differing technologies and properties are exploited to provide precise tracking as well as fast trigger information. Both in the barrel as well in the end-caps, the chambers are arranged in three layers: in the barrel the layers are arranged in concentric cylinders while in the endcaps in large wheels perpendicular to the beam pipe.

Precise measurements of the track coordinates are provided by two sub-detectors: at small η values ($|\eta| < 2.7$) there are the Monitored Drift Tubes (MDTs), while at large pseudorapidity values ($2 < |\eta| < 2.7$) there are the Cathode Strip Chambers (CSCs). The MDT chambers are aluminum drift tubes with a central tungstenium wire, filled with an argon gas. The CSCs are multiwire proportional chambers with cathodes segmented into strips. These chambers are characterized by high granularity and they are used in the innermost tracking layer due to their higher rate capability and time resolution.

The muon trigger system consists of Resistive Plate Chambers (RPCs) in the barrel region and of Thin Gap Chambers (TGCs) in the end-cap regions. These sub-detectors are exploited to provide fast trigger signals (15–25 ns) for muons traversing the pseudorapidity region $|\eta| < 2.4$. The RPCs are used in the barrel region $|\eta| < 1.05$ and they are arranged in concentric cyliders around the beam pipe. Each RPC consists of two parallel electrode plates in which the gap is filled with a $C_2H_2F_4$-based gas-mixture. The TGCs cover the region $1.05 < |\eta| < 2.4$. Each TGC module consists of a wire plane maintained at high positive voltage, sandwiched between resistive grounded cathode planes. The TGCs are well suited for triggering the high transverse momentum muons as they operate in a saturated proportional mode.

The MS is designed to offer a high transverse momentum resolution for muons. The resolution varies between 3 and 12% for p_T values between 10 GeV and 1 TeV. The spectrometer can measure muon momenta with adequate momentum resolution and excellent charge identification in the range between 3 GeV and 3 TeV. At lower energy the resolution is improved by adding the information of the tracks reconstructed in the ID.

2.2.5 *Trigger and Data Acquisition*

The ATLAS Trigger and Data AcQuisition (TDAQ) system is organized in two parallel pipelines: a read-out system (blue part in Fig. 2.8), which reads and transmits the electrical information from the detectors, and a trigger system (green part in Fig. 2.8), which receives the detector information from the read-out and decides which information should be recorded. The TDAQ is designed to reduce the nominal input rate of 40 MHz to an output rate of 1.5 kHz selecting only potentially interesting events. Events are discarded due to limitations regarding the computer resources for readout, storage and offline processing of data. The trigger system is composted by two separate levels: the Level 1 (L1) and the Higher Level Trigger (HLT).

The L1 trigger is a hardware based trigger designed to reduce the event rate from 40 MHz to 100 kHz. This trigger makes an initial selection based on reduced-granularity information from a subset of detectors, i.e. the calorimeters and the MS. The L1 trigger decision algorithm allows for isolating one or more Regions-of-Interest (RoIs)

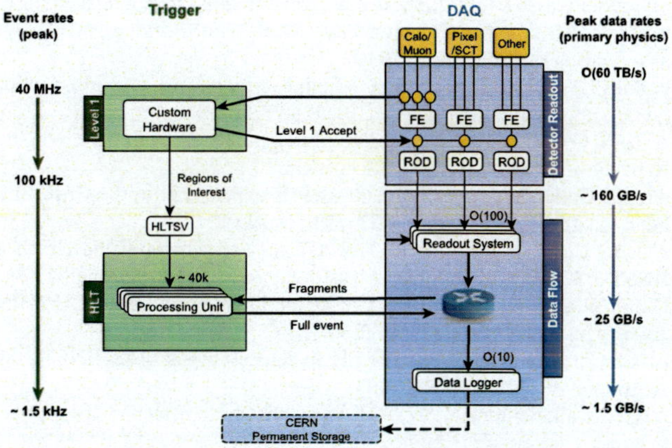

Fig. 2.8 Block diagram of the ATLAS trigger (left) and data acquisition (right) system [12]

which are the coordinates η and ϕ of the trigger sectors over threshold. In each RoIs, electron candidate objects, muon objects and $E_{\mathrm{T}}^{\mathrm{miss}}$ are reconstructed. The RoI data include information on the type of feature identified and the criteria passed. The decision of the L1 trigger is made in less than 2.5 μs. Memory pipelines are used to store information about subsequent trigger events. The L1 accepted events are merged into the Read-Out Drivers (ROD) and then sent to a computer farm implementing the second trigger stage. The RODs are detector-specific functional elements of the front-end systems, which achieve a higher level of data concentration and multiplexing by gathering information from several front-end data streams.

The HLT is a software based trigger designed to reduce the event rate from 100 to 1.5 kHz. The HLT refines the decision algorithm based on additional selection criteria. The latency for the HLT system is about 550 ms, a long time interval that allows for applying tighter selection criteria by using the full detector information. The HLT uses a farm of computer to reconstruct the events and it exploits algorithms very close to offline algorithms.

The configuration of the trigger system is done via a collection of *trigger chains* that specify the reconstruction sequence and selection steps from L1 to HLT. By changing the selection criteria and thresholds applied to the objects, the trigger rates can be controlled. In order to ensure that the HLT output rate is limited to 1.5 kHz, threshold of triggers in the menu are either increased or the triggers are *prescaled* by a factor N, meaning that only one in N events passing the selection rate is accepted for readout.

References

1. Breskin A, Voss R (2009) The CERN large hadron collider: accelerator and experiments, CERN. https://cds.cern.ch/record/1244506
2. Myers S, Picasso E (1990) The design, construction and commissioning of the CERN large electron positron collider. Contemp Phys 31:387. https://doi.org/10.1080/00107519008213789. https://cds.cern.ch/record/220034
3. Perkins DH (2000) Introduction to high energy physics, 4th edn. Cambridge University Press (2000). https://doi.org/10.1017/CBO9780511809040. https://cds.cern.ch/record/396126
4. ATLAS Collaboration, Run 2 luminosity public plots. https://twiki.cern.ch/twiki/bin/view/AtlasPublic/LuminosityPublicResultsRun2
5. ALICE Collaboration (2008) The ALICE experiment at the CERN LHC. JINST 3:S08002. https://doi.org/10.1088/1748-0221/3/08/S08002
6. ATLAS Collaboration (2008) The ATLAS experiment at the CERN large hadron collider. JINST 3:S08003. https://doi.org/10.1088/1748-0221/3/08/S08003
7. CMS Collaboration (2008) The CMS experiment at the CERN LHC. JINST 3:S08004. https://doi.org/10.1088/1748-0221/3/08/S08004
8. LHCb Collaboration (2008) The LHCb detector at the LHC. JINST 3:S08005. https://doi.org/10.1088/1748-0221/3/08/S08005
9. ATLAS Collaboration (1997) ATLAS magnet system: technical design report, 1, ATLAS-TDR-6. https://cds.cern.ch/record/338080
10. Potamianos K (2016) The upgraded pixel detector and the commissioning of the inner detector tracking of the ATLAS experiment for Run-2 at the large hadron collider, technical report.

ATL-PHYS-PROC-2016-104, p 15, EPS-HEP 2015 Proceedings: CERN. https://cds.cern.ch/record/2209070
11. ATLAS Collaboration (2010) ATLAS insertable B-layer technical design report, ATLAS-TDR-19. https://cds.cern.ch/record/1291633. Addendum: ATLAS-TDR-19-ADD-1, 2012, https://cds.cern.ch/record/1451888
12. ATLAS Collaboration, Trigger and data aquisition public plots. https://twiki.cern.ch/twiki/bin/view/AtlasPublic/ApprovedPlotsDAQ

Chapter 3
Dataset and Simulated Event Samples

This chapter outlines the dataset used by the analysis described in this thesis and the trigger strategy adopted to collect data. The signal and background processes contributing to the analysis are described as well, together with the Monte Carlo generators used to model them.

3.1 Dataset and Trigger

The data used in the analysis described in this thesis were collected at a centre-of-mass energy $\sqrt{s} = 13$ TeV during the 2015–2018 running period, corresponding to an integrated luminosity of 139 fb^{-1}. During the Run 2, the mean number of interactions per bunch crossing $< \mu >$ is about 34. Events used in data analysis are selected only if all the relevant systems of the ATLAS detector are known to be in good operation condition.

In the analysis the events are categorized in the 0-lepton, 1-lepton and 2-lepton channels, aiming at selecting $ZH \rightarrow \nu\nu b\bar{b}$, $WH \rightarrow l\nu b\bar{b}$ and $ZH \rightarrow llb\bar{b}$ events. The trigger algorithms used to select such events exploit the signatures of the leptonic decay of the V boson, depending on the leptonic channel. Events in the 0-lepton channel are selected using a missing transverse momentum $E_{\mathrm{T}}^{\mathrm{miss}}$ trigger because the events are characterized by the presence of two neutrinos from the Z boson decay. Selecting events that contain invisible particles is particularly difficult because such particles are not register in the detector. The strategy employed is to deduce the presence of invisible particles from the missing transverse momentum calculated from visible particles using only the calorimeter information.

In 1- and 2-lepton channels different triggers are used to select electron and muon decays of the W and Z bosons, respectively. In the electron sub-channel, a low-p_{T} threshold unprescaled single electron trigger is used, while in the muon sub-channel events are recorded using the same $E_{\mathrm{T}}^{\mathrm{miss}}$ triggers used in the 0-lepton channel. Muons are approximately invisible in the calorimeter so they contribute only marginally (i.e. they are treated like neutrinos) in the $E_{\mathrm{T}}^{\mathrm{miss}}$ calculations. The choice of using the $E_{\mathrm{T}}^{\mathrm{miss}}$

G. Di Gregorio, *A Beauty-ful Boson*, Springer Theses,
https://doi.org/10.1007/978-3-031-20013-7_3

trigger is motivated by the low muon trigger efficiency for $|\eta| > 2.4$ region due to the TGC coverage and some gaps needed for the inner detector services. Additionally the $E_{\mathrm{T}}^{\mathrm{miss}}$ triggers effectively select muons since their contribution is not included in the online $E_{\mathrm{T}}^{\mathrm{miss}}$ calculation. In the 1-lepton muon sub-channel, the $E_{\mathrm{T}}^{\mathrm{miss}}$ trigger provides an efficiency of 98% for signal events passing the offline selection, compared to a ~80% efficiency of the single-muon trigger. In the 2-lepton muon sub-channel, the $E_{\mathrm{T}}^{\mathrm{miss}}$ trigger is preferred with respect to the single muon one because it is more efficient (~5%).

3.1.1 E_T^{miss} *Trigger*

The $E_{\mathrm{T}}^{\mathrm{miss}}$ trigger algorithms used by ATLAS are based on the transverse momentum imbalance evaluated using only the calorimeter information. The first selection to build the $E_{\mathrm{T}}^{\mathrm{miss}}$ trigger algorithm is done at the L1 trigger level. Signals from calorimeter cells are used to form projective towers with granularity of $\Delta\eta \times \Delta\phi = 0.1 \times 0.1$ [1]. Fixed threshold is then applied per tower: the energy E_i of tower i below this threshold is set to zero in the subsequent calculations. All the towers over threshold are summed into larger towers with granularity of $\Delta\eta \times \Delta\phi = 0.2 \times 0.2$. The $E_{\mathrm{T}}^{\mathrm{miss}}$ is then recalculated doing a vectorial sum of the contributions of the clustered towers. Events that are accepted by the L1 trigger are then transferred to the HLT where the $E_{\mathrm{T}}^{\mathrm{miss}}$ is recalculated.

At HLT level $E_{\mathrm{T}}^{\mathrm{miss}}$ is calculated using jets reconstructed in the HLT with the anti-k_t algorithm with $R = 0.4$. During the Run 2 period the $E_{\mathrm{T}}^{\mathrm{miss}}$ algorithm used in HLT has been modified to remove the pile-up effects and to improve the performance.

The efficiency of the L1 $E_{\mathrm{T}}^{\mathrm{miss}}$ trigger is measured using $Z \to \mu\mu$ events selected with a muon trigger. The muons have a little interaction with the calorimeter, so the transverse momentum of the dimuon system $p_{\mathrm{T}}(\mu\mu)$ provides a good proxy of the expected $E_{\mathrm{T}}^{\mathrm{miss}}$. Figure 3.1a shows the L1 trigger efficiency as a function of $p_{\mathrm{T}}(\mu\mu)$ for a L1 nominal threshold of 50 GeV. The algorithm achieves 90% efficiency for $p_{\mathrm{T}}(\mu\mu)$ ~150 GeV and, at $p_{\mathrm{T}}(\mu\mu) > 200$ GeV it becomes fully efficient. Moreover it is observed that the efficiency is quite stable across the different $< \mu >$ values.

At the HLT level, the algorithm efficiency depends upon luminosity which has caused the increase of pile-up levels. The overall effect of the increase of the pile-up brings the degradation of the $E_{\mathrm{T}}^{\mathrm{miss}}$ resolution of the detector. To control the trigger rate due to the increasing level of pile-up, the HLT threshold was raised progressively from 70 GeV to 110 GeV during the Run 2. Figure 3.1b shows the overall (L1+HLT) $E_{\mathrm{T}}^{\mathrm{miss}}$ trigger efficiency for $Z \to \mu\mu$ events year-by-year as a function of $p_{\mathrm{T}}(\mu\mu)$. Due to the threshold increase, the efficiency remains quite stable within few percent in the full $p_{\mathrm{T}}(\mu\mu)$ range over the four years.

The trigger efficiency is 85–90% efficient at $E_{\mathrm{T}}^{\mathrm{miss}}$ =150 GeV and fully efficient above 180 GeV. The same values are obtained on simulated data applying the offline selection. Since the analysis described in this thesis exploits a phase space with $E_{\mathrm{T}}^{\mathrm{miss}}$

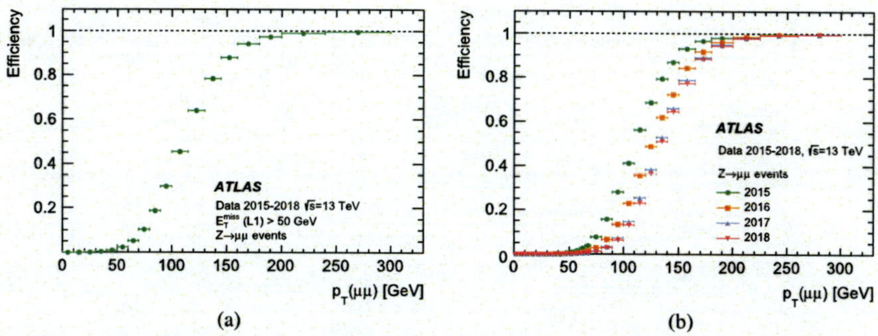

Fig. 3.1 **a** The L1 $E_{\mathrm{T}}^{\mathrm{miss}}$ trigger efficiency as a function of $p_{\mathrm{T}}(\mu\mu)$. **b** Full chain $E_{\mathrm{T}}^{\mathrm{miss}}$ trigger efficiencies for each year as a function of $p_{\mathrm{T}}(\mu\mu)$. In both plots the error bands correspond only to the statistical uncertainty [1]

$\geq$ 250 GeV and the $E_{\mathrm{T}}^{\mathrm{miss}}$ trigger is completely efficient above 180 GeV, there is no need to apply data-to-MC scale factors to correct $E_{\mathrm{T}}^{\mathrm{miss}}$ trigger inefficiencies.

3.1.2 *Single Electron Trigger*

The single electron triggers select events with electrons from the huge amount of data produced by high energy *p*-*p* collisions at the LHC and only a fraction of these events can be recorded.

The L1 trigger for electrons uses calorimeter information to build an EM RoI consisting of 4×4 trigger towers with a granularity of $\Delta\eta \times \Delta\phi = 0.1 \times 0.1$. The energy of the trigger towers is calibrated at the electromagnetic energy scale and a nominal transverse energy threshold is applied. The threshold is η-dependent to account for energy losses and the geometry of the detector. Additionally, EM isolation requirements are applied. The isolation requirements are optimized to maintain a fixed L1 efficiency at the lowest possible rate. At the HLT, the full detector granularity is used within the RoI for the final trigger decision. Electrons are identified with EM clusters and with tracks matched to the clusters. As in the offline algorithm, the electron selection relies on a multivariate technique using a likelihood-based discriminant with four operating points. The working point used in the trigger is designed to be as close as possible to the offline version. Isolation requirements are also added at this stage.

The electron trigger efficiency is measured for electrons at the HLT using the tag-and-probe method [2]. The trigger efficiency is defined as the ratio of the number of triggered electron candidates to the number of produced electrons and it is evaluated considering $Z \rightarrow ee$ events. The evolution of the single-electron trigger efficiency in 2015–2018 is shown in Fig. 3.2a. The sharper efficiency turn-on as a function of the transverse energy E_{T} in 2015 is due to a looser identification requirement, a lower E_{T}

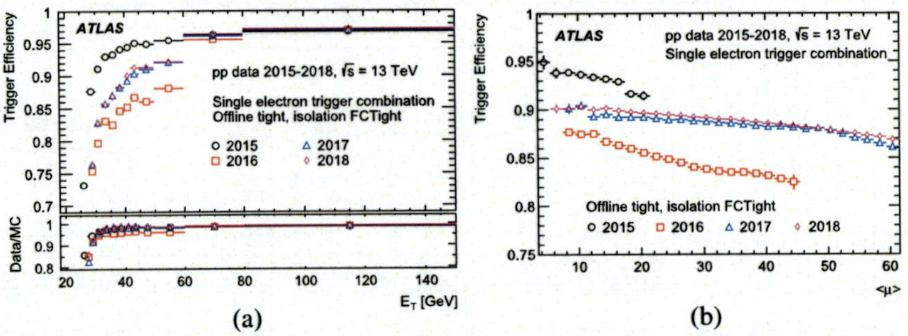

Fig. 3.2 **a** Evolution of the single-electron trigger efficiency as a function of the offline E_T during Run 2. In the bottom panel the ratios of data to MC simulation efficiencies are also shown. **b** Evolution of the single-electron efficiency as a function of pile-up during Run 2 [3]

threshold and no isolation requirement. Moreover 2016 data show some inefficiency at $E_\mathrm{T} < 60$ GeV due to too stringent online identification criteria.

Simulated events need to be corrected to reproduce as closely as possible the efficiencies measured in data. This is achieved applying scale factors defined as the ratio of the efficiency measurements in data to that one determined in simulated events. The data-to-MC scale factors, shown in the lower panel of Fig. 3.2a, are of the order of 18% close to the trigger E_T threshold and of the order of 4% above 40 GeV. The precision of these scale factors is of the order of 0.1%. Figure 3.2b shows the dependence of the trigger efficiency on pile-up, which is less pronounced at the second half of Run 2.

3.2 Simulated Samples

Simulated samples of signal and background processes are produced and studied for the optimization of the analysis criteria and to determine the expected signal and background distributions in the different regions of the analysis. Furthermore, simulations are also used to estimate the systematic uncertainties caused by different sources. The QCD multi-jet process is treated differently because the production cross-section is large (almost eight order of magnitude larger than the Higgs cross-section) and the rejection power of such events by the analysis selection is high. Such background is estimated using data-driven techniques, as described in Sect. 6.3.7. Simulated events are obtained with a multiple step process. The samples are produced with state-of-art Monte Carlo event generators. The simulation begins with a hard process in which the constituents of the colliding particles interact at high momentum scale to produce outgoing fundamental objects as the SM quarks, leptons, bosons

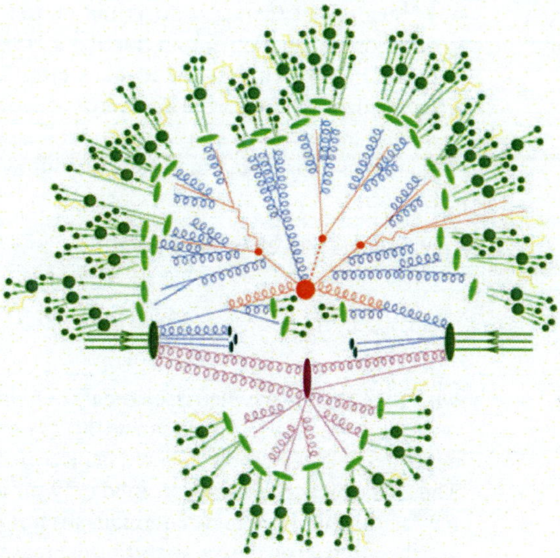

Fig. 3.3 Schematic diagram of the steps of the event simulation. The simulation of the hard process is shown in red while the parton shower is represented with blue lines. The production of hadrons and their decay is represented by green lines and dots. The purple lines represent the simulation of the underlying events [5]

or some new particles of BSM theories. The partons from the hard process radiate gluons as they evolve, leading to the formation of parton shower. The constituents of the parton showers hadronise to form colourless hadrons triggering the process of hadronisation. Many of the produced hadrons are unstable, so another step of the simulation is the hadron decay. In hadron-hadron collisions, the other constituent partons of the incoming hadrons undergo multiple interactions which produce the underlying events. The final step is a detailed simulation of the ATLAS detector response based on Geant 4 [4]. All the simulation steps, except for the detector simulation response, are schematically shown in Fig. 3.3 and briefly summarised in the following.

- **Hard process simulation:** Many LHC processes involve large momentum transfers such the production of jets with high transverse momenta. The simulation of these processes is the core of any simulation of collider events through Monte Carlo generators. Such interactions can be described by perturbation theory since QCD quanta are asymptotically free. The cross-section for a scattering process of two hadrons $h1$ and $h2$ at hadron colliders can be computed as:

$$\sigma_{h1h2\rightarrow n} = \sum_{a,b} \int_0^1 \int_0^1 dx_a dx_b \int f_a^{h1}(x_a, \mu_F) f_b^{h2}(x_b, \mu_F) d\hat{\sigma}_{ab\rightarrow n}(\mu_F, \mu_R) \tag{3.1}$$

where:

- $f_a^{h1}(x_a, \mu_F)$ ($f_b^{h2}(x_b, \mu_F)$) is the parton distribution function[1] (PDF) which depends on the momentum fraction x_a (x_b) of a parton a (b) with respect to its parent hadron $h1$ ($h2$) and on the factorization scale μ_F;
- $\hat{\sigma}_{ab\to n}$ denotes the parton-level cross-section for the production of the final state n from the initial partons a and b which depends on the momenta given by the final phase space Φ_n, on the factorization scale μ_F and on the renormalisation scale μ_R.

Equation 3.1 can be re-written as:

$$\sigma = \sum_{a,b} \int_0^1 \int_0^1 dx_a dx_b \int d\Phi_n f_a^{h1}(x_a, \mu_F) f_b^{h2}(x_b, \mu_F) \times \frac{1}{2\hat{s}} |\mathcal{M}_{ab\to n}(\Phi_n; \mu_F, \mu_R)|^2 \tag{3.2}$$

where $\mathcal{M}_{ab\to n}$ is the matrix element, $d\Phi_n$ is the differential phase space element over the n final-state particles and $1/(2\hat{s}) = 1/(2x_a x_b s)$ is the parton flux which depends on the hadronic squared centre-of-mass energy s.
The cross-section is fully specified only for a given PDF set and a given choice of the factorization μ_F and renormalisation μ_R scales. These scales are unphysical and their impact on the cross-section decreases in higher order terms of perturbation theory which are taken into account.

- **Parton shower:** In the event simulation, the hard process is described with the lowest-order matrix elements. The higher orders of the process can be approximate through a parton shower algorithm, which is formulated as an evolution in momentum transfer from the high energy scales, associated to the hard process, to the low scales (of the order of 1 GeV), associated to the confinement of partons into hadrons. The parton shower represents the connection between the scattered partons and the final hadrons, and it aims at modelling this extra radiation before hadronisation.
- **Hadronisation:** As the momentum decreases perturbation theory can not be used any more to describe the QCD interactions. At this stage, the perturbative evolution must be replaced by a non-perturbative hadronisation model. This model describes how the quarks and gluons from hard scattering simulation, parton showers and multiple scattering simulations are transformed into colorless final states. Individual parton hadronisation is not allowed, instead color-connected systems of partons hadronise collectively.
- **Decays:** Following the hadronisation phase of the event generation, unstable hadrons which decay into lighter hadrons are produced. Unstable hadrons decay into particles that are considered stable on the detector timescales. This is an important part of the event simulation because the observed final-state hadrons result from a convolution of hadronisation and decay processes.
- **Underlying events:** The underlying event simulation contains all the events not coming from the primary hard scattering process. The underlying event includes

[1] The parton distribution function is used to described the probability to find a parton carrying a fraction x of the total momentum. Different parton distribution functions are realised by several collaborations.

contribution from initial and final state radiation, beam-beam remnants or multiple parton interactions. The energy scale of the underlying process is expected to be smaller than the one of the hard scattering event, the underlying event has a uniform activity represented by low energy hadrons. The simulation of pile-up is an independent step of the simulation. This means that the simulation of the hard scattering event is overlaid with these simulated pile-up events.

- **Pile-up events:** At LHC there is high chance that multiple inelastic interactions in the same bunch (in-time pile-up) and neighbouring bunch crossing (out-of-time pile-up) happen simultaneously at the collision point. Each component of the pile-up has a dedicated simulation. The pile-up simulation is achieved by overlaying simulated minimum bias events to signal events prior to the conversion of energy deposits to detector signals [6].
- **Detector simulation:** The final step in the Monte Carlo chain is the simulation of interactions between the final state particles created by the event generators and the ATLAS detector. The detector simulation describes particle-matter interactions, particles trajectories and particle decays within the ATLAS detector volume. This simulation is based on Geant 4 and events are reconstructed with the ATLAS reconstruction software [7]. The detector simulation follows the detector state evolutions in term of active subdetectors, dead channels and misalignment. The configuration of the detector, including the misalignment and distortions, can be set at run time. The energies deposited in the active portions of the detector are recorded as *hits*, containing the total energy deposition, position and time, and then they are written to a simulation output file. The detector simulation is quite slower than the event generation, and for this reason it is convenient to simulate the interaction with the ATLAS detector only if the events pass the full analysis chain. *MC filters* are applied between the generation and simulation steps to select events in a particular phase space.

3.2.1 Monte Carlo Generators

Several programs are available to simulate the events. There are some multi-purpose generators and some specialized programs. The former type of generator is able to simulate the full event from the matrix element to the hadronisation, while the latter type generates only part of the event. In this case an interface for the specialized programs is needed. The MC event generators used in the analysis are described in the following.

- **`Pythia`:** `Pythia` [8] is a multi-purpose generator providing parton shower, underlying event and matrix element calculations. The latter calculation is given at leading order (LO), while both parton shower and underlying event models are tuned on existing measurements. Several PDFs are hardcoded in this event generator. This generator can be used to simulate just certain parts of the event. This feature is useful since the accuracy of the matrix element calculation is just at LO. However, the parton shower model exhibits good agreement with data therefore Pythia is often used to provide the parton shower description for a simulation model.

- **Herwig:** Herwig [9] is a multi-purpose particle physics event generator but it is not used as a stand-alone generator any more. However it is still used for the parton shower description of a simulation model. Many simulations that use Pythia to describe the parton shower are also generated with Herwig, the comparison between the two generators allows to evaluate the systematic effects of the parton shower model on the simulation.
- **Sherpa:** Sherpa [10] is a general-purpose event generator, capable of simulating the physics of hadron-hadron collisions. It covers both the matrix element calculation and the parton shower description. It provides next-to-leading order (NLO) calculation for a variety of processes. It is the preferred generator for processes with additional radiated jets, which are directly included in the matrix-element calculation. Differently from Pythia and Herwig, there are no "hybrid" versions of Sherpa, therefore Sherpa can not be used only to generate part of the event.
- **Powheg:** Powheg [11] is a generator able to simulate the matrix elements with NLO calculation for a variety of processes. It is used together with other generators that simulate the parton shower.
- **MadGraph:** MadGraph [12] is a generator that provides only the matrix element calculations at LO and NLO. In most of the case it is not used as default matrix element generator but it is used to study systematic effects.

All ATLAS simulated dataset model pile-up events with Pythia 8.1 [8]. In addition, all simulations, except Sherpa, use the EvtGen [13] program to describe the decay of hadrons containing bottom or charm quarks.

3.3 Signal Process Simulation

The VH signals studied in this thesis include three main Higgs processes: $ZH \rightarrow \nu\nu b\bar{b}$, $WH \rightarrow l\nu b\bar{b}$ and $ZH \rightarrow llb\bar{b}$ where $l = e, \mu, \tau$. The tree-level Feynman diagrams of the three different processes are shown in Fig. 3.4.

The ZH production mode is furthermore split into two contributions depending on the partonic initial state, $qq \rightarrow ZH$ or $gg \rightarrow ZH$. Two LO Feynman diagrams for the $gg \rightarrow ZH$ process are shown in Fig. 3.5.

All the qq-initiated production processes are simulated using the Powheg generator with the MiNLO (Multiscale Improved Next-to-Leading Order) procedure [14] applying AZNLO tune [15] with NNPDF3.0NLO PDF set [16] for the matrix-element. The events are also interfaced to Pythia 8.212 [17] MC generator for the modelling of the parton shower, underlying event and multiple parton interactions. The simulation of qq-initiated production processes includes also qg-initiated

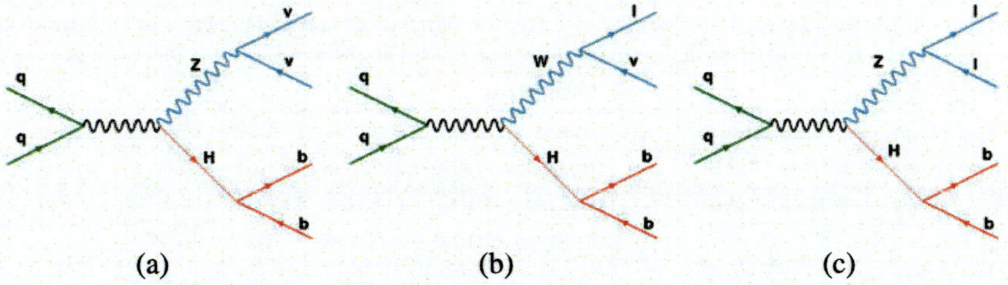

Fig. 3.4 Leading order Feynman diagram for $ZH \to \nu\nu b\bar{b}$, $WH \to l\nu b\bar{b}$ and $ZH \to llb\bar{b}$ processes with $l = e, \mu, \tau$. These diagrams represent the 0-, 1-, and 2-lepton channels

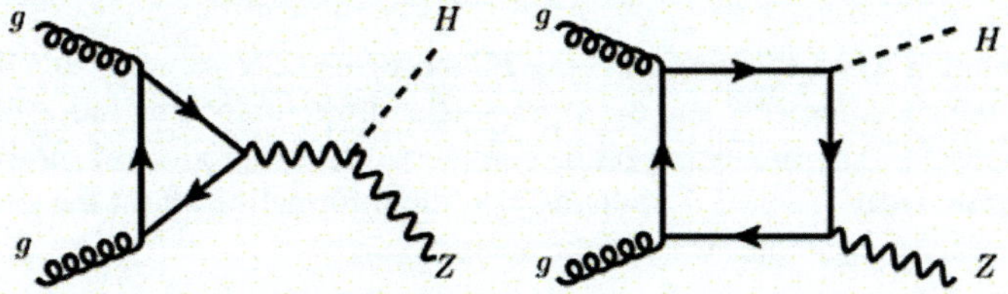

Fig. 3.5 Two leading order Feynman diagrams of the $gg \to ZH$-process

Fig. 3.6 One of the leading order Feynman diagram of the qg-initiated process

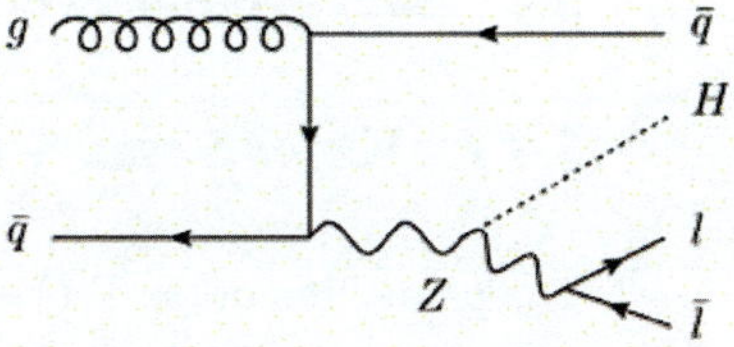

events. An example of a Feynman diagram of the qg-initiated process is shown in Fig. 3.6.

The gg-initiated events are simulated using the `Powheg` generator using NNPDF3.0NLO PDF set for the matrix-element. Furthermore the events are interfaced to `Pythia 8.212` generator for the modelling of parton shower, underlying event and multiple parton interactions. The gg-initiated events are generated at LO with NLO corrections.

All the samples are normalised to the best available theoretical prediction for the cross-section of each process at $\sqrt{s} = 13$ TeV. The signal cross-sections are evaluated following the prescriptions of the LHC Higgs Cross-Section Working Group for a Higgs boson mass of $m_H = 125$ GeV [18, 19], considering the $b\bar{b}$ branching ratio fixed to 58%. The WH signal samples are normalised to the production cross-section calculated at next-to-next-to-leading order (NNLO) in QCD with electroweak (EW) corrections at NLO accuracy. The $gg \to ZH$ cross-section is calculated at NLO in QCD, while for the $qq \to ZH$ process the cross-section is evaluated from the difference between the total ZH cross-section, calculated at NNLO in QCD with NLO EW corrections, and the gluon-induced ZH production. This is done to avoid double

Table 3.1 Monte Carlo samples used for the signal processes and the cross-section times branching ratio (BR) used to normalise the different processes at $\sqrt{s} = 13$ TeV l is inclusive of e, μ, τ leptons

Process	Generator	$\sigma\times$ BR [fb]
$qq \rightarrow ZH \rightarrow \nu\nu b\bar{b}$	`Powheg MiNLO` + `Pythia 8.212`	89.08
$qq \rightarrow WH \rightarrow l^{+}\nu b\bar{b}$	`Powheg MiNLO` + `Pythia 8.212`	164.6
$qq \rightarrow WH \rightarrow l^{-}\nu b\bar{b}$	`Powheg MiNLO` + `Pythia 8.212`	104.5
$qq \rightarrow ZH \rightarrow llb\bar{b}$	`Powheg MiNLO` + `Pythia 8.212`	44.84
$gg \rightarrow ZH \rightarrow \nu\nu b\bar{b}$	`Powheg` + `Pythia 8.212`	14.30
$gg \rightarrow ZH \rightarrow l^{-}l^{+}b\bar{b}$	`Powheg` + `Pythia 8.212`	7.23

counting when merging $qq \rightarrow ZH$ and $gg \rightarrow ZH$ samples together. A summary of the signal process and the cross-section times branching ratio values are listed in Table 3.1. The branching ratios correspond to the product of the Higgs and vector boson decay. Table 3.1 contains also the information about the generators used to produce the samples used in the analysis.

3.4 Background Process Simulation

3.4.1 Vector Boson + Jets Production

The production of the V ($V = Z$ or W) boson in association with jets is one of the main backgrounds in all the lepton channels considered in the analysis. Figure 3.7 shows the lowest order Feynman diagrams for Z+jets and W+jets background.

The production of high statistic V+jets MC samples is important for this and many other ATLAS analyses. Efforts have been made by the ATLAS Collaboration to update the simulations to the latest V+jets MC generators, moving from LO to NLO generator, and updating parameter tunings for the Run 2 data taking [20]. The V+jets

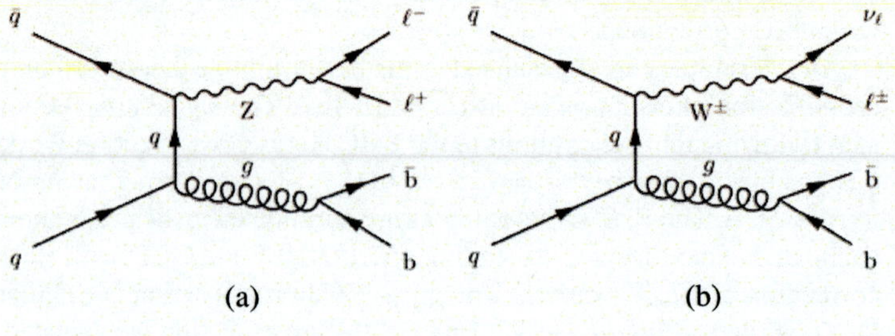

Fig. 3.7 Lowest order Feynman diagrams for the quark induced Z+jets (**a**) and W+jets (**b**) processes

Table 3.2 Monte Carlo samples used for the V+jets processes and the cross-section times branching ratio used to normalise the different processes at $\sqrt{s} = 13$ TeV. l is inclusive of e, μ, τ leptons

Process	Generator	$\sigma \times$BR [pb]
$Z(\rightarrow \nu\nu)$ + jets	Sherpa 2.2.1	1914
$W(\rightarrow l\nu)$ + jets	Sherpa 2.2.1	20080
$Z/\gamma^{*}(\rightarrow ll)$ + jets [$m_{ll} > 40$ GeV]	Sherpa 2.2.1	2107

processes are simulated with `Sherpa 2.2.1` using NNPDF3.0NNLO PDF set [16] with dedicated parton shower tuning developed by `Sherpa` authors. Events with many jets produced in association with the W or the Z boson largely contribute to the background in the phase space of the analysis. `Sherpa 2.2.1` provides a combination of different matrix elements with different parton multiplicities. Events with zero or one additional partons are generated at NLO in the ME calculation, while events with two or three additional partons are included at LO in the ME. Samples are normalised using cross-sections calculated at NNLO accuracy [21]. The list of the V+jets samples used and the relative cross-sections is given in Table 3.2.

The analysis treated in this thesis includes different regions of phase space corresponding to high p_{T} of the vector boson (p_{T}^{V}). To generate sufficient high p_{T}^{V} events, the V+jets samples are split depending on p_{T}^{V} and the scalar p_{T} sum of all parton-level jets with $p_{\mathrm{T}} > 20$ GeV called H_T. Samples are produced in slices of $\max(H_T, p_{\mathrm{T}}^{V})$ according to the following intervals:

$$[0 - 70, 70 - 140, 140 - 280, 280 - 500, 500 - 1000, > 1000] \,\text{GeV}$$

Final states with b-tagged jets are studied in this analysis. It is important to make sure that the MC statistics in this specific heavy-flavour enriched phase space are large enough to provide a robust MC prediction to be compared to the data. For this reason the V+jets samples are generated applying different filters to select the flavour composition of the jets produced in association with the vector boson V. Three filters are used:

- *BFilter*: at least a b-hadron with $|\eta|$ <4 is required;
- *CFilterBVeto*: at least a c-hadron with p_{T} >4 GeV and $|\eta|$ <3 is required, and no overlap with the BFilter sample;
- *CVetoBVeto*: events which pass the previous two filters are rejected.

3.4.2 Top Pair Production

The top pair process is a main background for both the 0-lepton and 1-lepton channels. Figure 3.8 shows an example of lowest order Feynman diagrams for the top pair background.

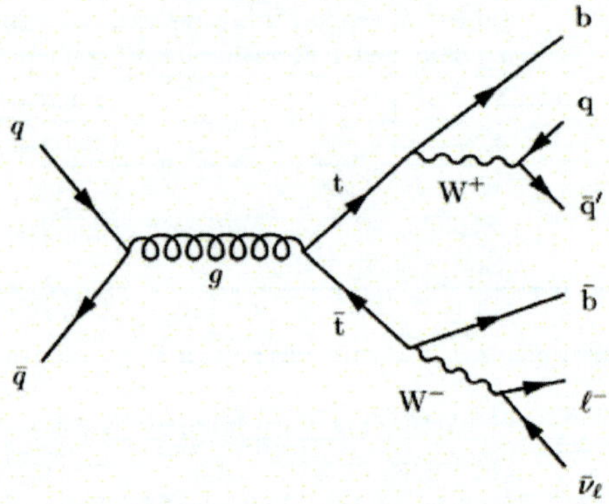

Fig. 3.8 Lowest order Feynman diagram for the $t\bar{t}$ process

The MC configuration for the $t\bar{t}$ process is `Powheg + Pythia 8.230`, which utilities the `Powheg` NLO matrix element generator [22, 23] interfaced to `Pythia 8.230` using the A14 tune [24] for the parton shower, hadronisation, underlying event and multiple parton interactions. Additionally the sample uses NNPDF3.0NLO PDF set in the matrix element calculation. The cross-section used to normalised the $t\bar{t}$ sample is calculated at NNLO in QCD including resummation of next-to-next-to-leading logarithmic (NNLL) soft gluon terms [25]. Table 3.3 briefly summaries the generators used for the top pair production together with the cross-section value. To have enough statistics, the samples used in the $VH(b\bar{b})$ analysis are filtered at generator stage using $E_{\mathrm{T}}^{\mathrm{miss}}$ filters.[2]

3.4.3 Single-Top Production

The single-top process is a relevant background for both 0-lepton and 1-lepton channels. Three different channels are generated: s-channel, t-channel and W-associated production (Wt-channel). Among the three channels, Wt-channel is the one which gives the most relevant contribution in the analysis. Figure 3.9 shows the lowest order Feynman diagrams for the three channels of single-top production.

The single-top background events are generated separately for the different channels, using the `Powheg` generator for the hard scattering process with the

Table 3.3 Monte Carlo samples used for the $t\bar{t}$ process and the cross-section times branching ratio (BR) used to normalise the different processes at $\sqrt{s} = 13$ TeV

Process	Generator	$\sigma \times$ BR [pb]
$t\bar{t}$	`Powheg + Pythia 8.230`	831.76

[2] The $E_{\mathrm{T}}^{\mathrm{miss}}$ values used for the different slices are defined using the truth level information.

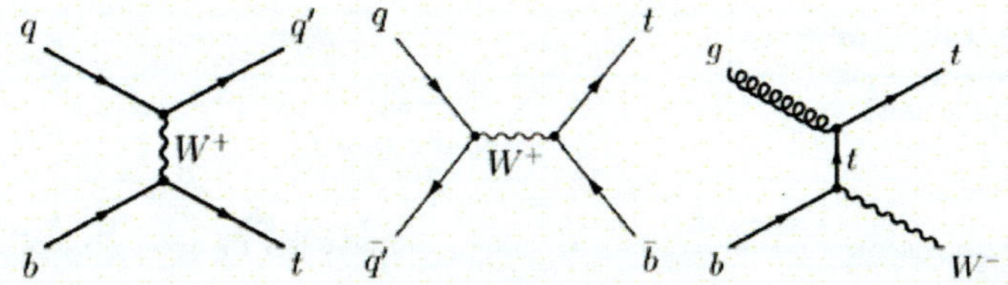

Fig. 3.9 Lowest order Feynman diagrams for the three channels of the single top production: t-channel (left), s-channel (middle), Wt-channel (right)

NNPDF3.0NLO PDF set. The events are generated at NLO in the ME calculation and interfaced with `Pythia 8.230` algorithm for the simulation of parton shower, underlying event and multiple parton interactions. Single top samples in t- and s-channels are generated applying a lepton filter to require the leptonic decay of the W boson, while the Wt-channel samples are generated applying either a dilepton filter (both W bosons decay leptonically) or without applying any filter. In the analysis, both Wt-channel samples are used, after removing the overlap between the two samples.

The predicted t-channel single-top cross-section for p-p collisions at center-of-mass energy $\sqrt{s} = 13$ TeV is $\sigma_t = 136.02^{+5.4}_{-4.6}$ pb for the top quark and $\sigma_{\bar{t}} = 80.95^{+4.1}_{-3.6}$ pb for the anti-top quark. In the same condition the s-channel single-top cross-section is $\sigma_t = 6.35^{+0.2}_{-0.2}$ pb for the top quark and $\sigma_{\bar{t}} = 3.97^{+0.19}_{-0.17}$ pb for the anti-top quark. The Wt-channel single-top cross-section is $\sigma_t = 71.7^{+3.8}_{-3.8}$ pb. The cross-sections of the t- and s-channels are calculated at NLO while the cross-section of the Wt-channel is computed at NNLO. In both cases, the top mass value used in the calculation is $m_t = 172.5$ GeV. All the single-top samples used in the analysis are normalised to the cross-sections from the higher order calculation. The full list of single-top samples used in the analysis, together with the generator information and the cross-section value times branching ratio, is shown in Table 3.4.

Table 3.4 Monte Carlo samples used for the single top processes and the cross-section times branching ratio (BR) used to normalise the different processes at $\sqrt{s} = 13$ TeV

Process	Generator	$\sigma \times$ BR [nb]
Single top t-channel, t	`Powheg + Pythia 8.230`	0.0370
Single top t-channel, $\bar{t}$	`Powheg + Pythia 8.230`	0.0222
Single top s-channel, t	`Powheg + Pythia 8.230`	0.0020
Single top s-channel, $\bar{t}$	`Powheg + Pythia 8.230`	0.0013
Single top Wt-channel, t	`Powheg + Pythia 8.230`	0.0380
Single top Wt-channel, $\bar{t}$	`Powheg + Pythia 8.230`	0.0380

Fig. 3.10 Lowest order Feynman diagrams for the diboson process

3.4.4 Diboson Production

The production of two vector bosons is a sub-dominant background process but important because it has a final state similar to the one of the signal. The diboson background consists of final states generated by WW, WZ and ZZ events. Several diboson processes contribute in the analysis phase space: $ZZ \rightarrow b\bar{b}\nu\bar{\nu}$ in 0-lepton channel, $ZW \rightarrow b\bar{b}l\nu$ in 1-lepton channel and $ZZ \rightarrow b\bar{b}l\bar{l}$ in 2-lepton channel. Additional processes as $WW \rightarrow q\bar{q}l\nu$ can give a smaller contribution arising from the fakes reconstruction of one of the leptons or from the mistagging of a jet from the W decay. Figure 3.10 shows the leading order Feynman diagrams for the diboson production.

Quark-induced diboson processes are simulated using `Sherpa 2.2.1` generator for both the matrix element and parton shower calculation with the NNPDF3. 0NNLO PDF set. `Sherpa` provides a combination of different matrix elements with different parton multiplicities: processes with zero or one additional partons are

Table 3.5 Monte Carlo samples used for the diboson process and the cross-section times branching ratio (BR) used to normalise the different processes at $\sqrt{s} = 13$ TeV

Process	Generator	$\sigma \times$ BR [nb]
$qq \rightarrow W^+W^- \rightarrow l\nu qq$	`Sherpa 2.2.1`	24.710
$qq \rightarrow W^+W^- \rightarrow qql\nu$	`Sherpa 2.2.1`	24.73
$qq \rightarrow WZ \rightarrow l\nu qq$	`Sherpa 2.2.1`	11.4
$qq \rightarrow WZ \rightarrow qqll$	`Sherpa 2.2.1`	3.44
$qq \rightarrow WZ \rightarrow qq\nu\nu$	`Sherpa 2.2.1`	6.80
$qq \rightarrow ZZ \rightarrow qqll$	`Sherpa 2.2.1`	15.56
$qq \rightarrow WZ \rightarrow l\nu b\bar{b}$	`Sherpa 2.2.1`	2.50
$qq \rightarrow ZZ \rightarrow llbb$	`Sherpa 2.2.1`	15.54
$qq \rightarrow ZZ \rightarrow \nu\nu bb$	`Sherpa 2.2.1`	15.57
$qq \rightarrow ZZ \rightarrow qq\nu\nu$	`Sherpa 2.2.1`	15.56
$gg \rightarrow W^-W^+ \rightarrow l\nu qq$	`Sherpa 2.2.2`	0.6224
$gg \rightarrow W^+W^- \rightarrow l\nu qq$	`Sherpa 2.2.2`	0.6225
$gg \rightarrow ZZ \rightarrow llqq$	`Sherpa 2.2.2`	0.923
$gg \rightarrow ZZ \rightarrow \nu\nu qq$	`Sherpa 2.2.2`	0.925

generated at NLO in the matrix element, while two or three additional partons are generated at LO in the matrix element. Gluon-induced samples ($gg \rightarrow VV$, with $V = Z$ or W boson) are generated using Sherpa 2.2.2 with NNPDF3.0NNLO PDF set. In this case the gg-initiated samples are generated at LO in the matrix element.

All the samples are normalised to the NLO inclusive cross-section. The inclusive cross-sections for the various diboson processes are summarized in Table 3.5.

References

1. Aad G et al. (2020) Performance of the missing transverse momentum triggers for the ATLAS detector during Run-2 data taking. arXiv:2005.09554 [hep-ex]
2. ATLAS Collaboration (2017) Electron efficiency measurements with the ATLAS detector using 2012 LHC proton-proton collision data. Eur Phys J C 77:195. https://doi.org/10.1140/epjc/s10052-017-4756-2. arXiv:1612.01456 [hep-ex]
3. ATLAS Collaboration (2019) Performance of electron and photon triggers in ATLAS during LHC Run 2. arXiv:1909.00761 [hep-ex]
4. Agostinelli S et al. (2003) GEANT4 - a simulation toolkit. Nucl Instrum Meth A 506:250. https://doi.org/10.1016/S0168-9002(03)01368-8
5. Simulations. https://theory.slac.stanford.edu/our-research/simulations
6. Marshall Z (2014) Simulation of pile-up in the ATLAS experiment. J Phys Conf Ser 513. ed. by D. Groep and D. Bonacorsi 022024. https://doi.org/10.1088/1742-6596/513/2/022024
7. ATLAS Collaboration (2010) The ATLAS simulation infrastructure. Eur Phys J C 70:823. https://doi.org/10.1140/epjc/s10052-010-1429-9. arXiv:1005.4568 [physics.ins-det]
8. Sjostrand T, Mrenna S, Skands PZ (2008) A brief introduction to PYTHIA 8.1. Comput Phys Commun 178:852. https://doi.org/10.1016/j.cpc.2008.01.036. arXiv:0710.3820 [hep-ph]
9. Bellm J et al. (2016) Herwig 7.0/Herwig++ 3.0 release note. Eur Phys J C 76:196. https://doi.org/10.1140/epjc/s10052-016-4018-8. arXiv:1512.01178 [hep-ph]
10. Gleisberg T, Höche S, Krauss F, Schönherr M, Schumann S et al. (2009) Event generation with SHERPA 1.1. JHEP 02:007. https://doi.org/10.1088/1126-6708/2009/02/007. arXiv:0811.4622 [hep-ph]
11. Alioli S, Nason P, Oleari C, Re E (2010) A general framework for implementing NLO calculations in shower Monte Carlo programs: the POWHEG BOX. JHEP 06:043. https://doi.org/10.1007/JHEP06(2010)043. arXiv:1002.2581 [hep-ph]
12. Alwall J et al. (2014) The automated computation of tree-level and next-to-leading order differential cross sections, and their matching to parton shower simulations. JHEP 07:079. https://doi.org/10.1007/JHEP07(2014)079. arXiv:1405.0301 [hep-ph]
13. Lange DJ (2001) The EvtGen particle decay simulation package. Nucl Instrum Meth A 462:152. https://doi.org/10.1016/S0168-9002(01)00089-4
14. Luisoni G, Nason P, Oleari C, Tramontano F (2013) $HW^{\pm}$/HZ+ 0 and 1 jet at NLO with the POWHEG BOX interfaced to GoSam and their merging within MiNLO. JHEP 10:083. https://doi.org/10.1007/JHEP10(2013)083. arXiv:1306.2542 [hep-ph]
15. ATLAS Collaboration (2011) Measurement of the transverse momentum distribution of Z/λ^* bosons in proton-proton collisions at $\sqrt{s}$ = 7 TeV with the ATLAS detector. Phys Lett B 705:415. https://doi.org/10.1016/j.physletb.2011.10.018. arXiv:1107.2381 [hep-ex]
16. Ball RD et al. (2013) Parton distributions with LHC data. Nucl Phys B 867:244. doi: 10.1016/j.nuclphysb.2012.10.003. arXiv:1207.1303 [hep-ph]
17. Sjöstrand T et al. (2015) An introduction to PYTHIA 8.2. Comput Phys Commun 191:159. https://doi.org/10.1016/j.cpc.2015.01.024. arXiv:1410.3012 [hep-ph]
18. LHCSW Group, SM Higgs production cross sections at 13 TeV. https://twiki.cern.ch/twiki/bin/view/LHCPhysics/CERNYellowReportPageAt13TeV

19. de Florian D et al. (2016) Handbook of LHC Higgs cross sections: 4. Deciphering the nature of the Higgs sector, 2/2017. https://doi.org/10.23731/CYRM-2017-002. arXiv:1610.07922 [hep-ph]
20. ATLAS Collaboration (2016) Monte Carlo generators for the Production of a W or Z/λ* Boson in association with jets at ATLAS in Run 2, ATL-PHYS-PUB-2016-003. https://cds.cern.ch/record/2120133
21. Catani S, Cieri L, Ferrera G, de Florian D, Grazzini M (2009) Vector boson production at hadron colliders: a fully exclusive QCD calculation at NNLO. Phys Rev Lett 103:082001. https://doi.org/10.1103/PhysRevLett.103.082001.arXiv:0903.2120 [hep-ph]
22. Frixione S, Nason P, Ridolfi G (2007) A Positive-weight next-to-leading-order Monte Carlo for heavy flavour hadroproduction. JHEP 09:126. https://doi.org/10.1088/1126-6708/2007/09/126. arXiv:0707.3088 [hep-ph]
23. Nason P (2004) A new method for combining NLO QCD with shower Monte Carlo algorithms. JHEP 11:040. https://doi.org/10.1088/1126-6708/2004/11/040. arXiv:hep-ph/0409146
24. ATLAS Collaboration (2014) ATLAS Pythia 8 tunes to 7 TeV data, ATL-PHYS-PUB-2014-021. https://cds.cern.ch/record/1966419
25. Beneke M, Falgari P, Klein S, Schwinn C (2012) Hadronic top-quark pair production with NNLL threshold resummation. Nucl Phys B 855:695. https://doi.org/10.1016/j.nuclphysb.2011.10.021. arXiv:1109.1536 [hep-ph]

Chapter 4
Object Reconstruction

Each recorded event undergoes offline event reconstruction. The output information from all sub-detectors is combined to form basic quantities such as tracks and calorimetric clusters. This quantities are then used to reconstruct the final physics objects. This chapter describes the reconstruction and the identification of all the physics objects. The described procedures are applied to both simulated events and data samples.

4.1 Tracks and Primary Vertex

The reconstruction of tracks from hits[1] in the tracking detectors caused by crossing charged particles is an important ingredient to their reconstruction and identification. Tracks are reconstructed in the inner detector using a subsequence of algorithms [1]. The main ones are the *inside-out* and a consecutive *outside-in* track-reconstruction approaches. The former is the baseline algorithm designed for the efficient reconstruction of primary charged particles.[2] The track search starts from information coming from the pixel and silicon detectors and it adds hits moving away from the interaction point using a combinatorial Kalman filter [2]. The latter algorithms is designed to reconstruct secondaries, which are particles emerging from the decays of primaries and it uses an apposite approach starting from the information of the TRT.

The increasing detector occupancy with pile-up can result in nearby hits from other particles confusing the pattern recognition algorithm such that the track is not correctly reconstructed. Increased occupancy can lead to an increase in combinatorial

[1] A hit is a measurement point assigned to a track.

[2] Primary particles are defined as particles with a mean life of greater than $3 \cdot 10^{-11}$ s directly produced in a *p-p* interaction or from the subsequent decays or interaction of particles with a lifetime shorter than $3 \cdot 10^{-11}$ s.

G. Di Gregorio, *A Beauty-ful Boson*, Springer Theses,
https://doi.org/10.1007/978-3-031-20013-7_4

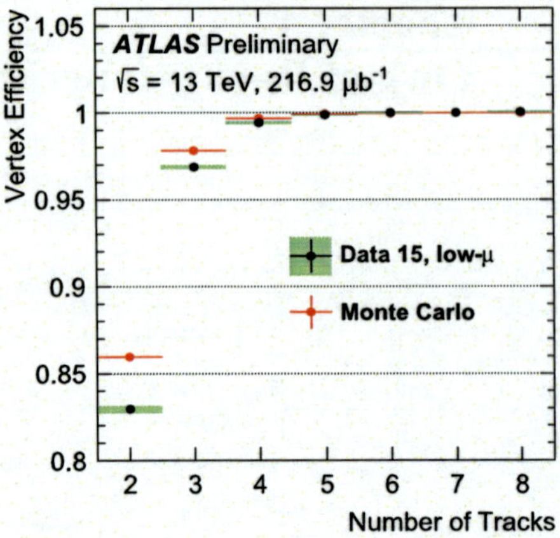

Fig. 4.1 Vertex reconstruction efficiency as a function of the number of tracks in low-μ data taken in 2016 compared to Monte Carlo simulation [4]

fake tracks which could not be matched to either a primary or a secondary vertex. Fake tracks can be minimize by tightening the quality requirements on reconstructed tracks. The selection criteria to the tracks require at least 9 hits in the silicon detectors (pixel + SCT), exactly zero holes[3] in the pixel detectors and a $p_{\mathrm{T}} > 400$ MeV.

Primary vertices are defined as the points in space where p-p interaction have occurred [3]. The primary vertex is of direct relevance to the reconstruction of hard-scattering interactions, in which the correct assignment of charged-particle trajectories to the hard-scattering primary vertex is used in reconstructing the full kinematic properties of the event. The input of the vertex reconstruction is a collection of reconstructed tracks. The selection of tracks likely stemming from the hard scatter vertex is mainly based on two variables: the transverse and the longitudinal impact parameters denoted d_0 and z_0, respectively. The transverse impact parameter d_0 is defined as the shortest distance between a track and the beam line in the transverse plane. The longitudinal impact parameter z_0 is defined as the distance in the z-axis between the primary vertex and the point on the track used to evaluate d_0. Cuts on the impact parameters are also used in the electron and muon candidate selection.

The procedure of primary vertex reconstruction is divided in two stages: vertex finding and vertex fitting. The former stage is the association of reconstructed tracks to vertex candidates, the latter stage deals with the reconstruction of the actual vertex position. After the determination of the vertex position, tracks that are incompatible with the vertex are removed from it and used in the determination of another vertex. The primary vertex is determined as the one with the highest sum of the transverse momenta squared (p_{T}^2) of all associated tracks. The remaining vertices are from pile-up and secondary vertices, which are fundamental for the b-jet identification. Figure 4.1 shows the vertex reconstruction efficiency as a function of the numbers of tracks [4]. The vertex reconstruction efficiency is evaluated from data considering

[3] A hole is a non-existing but expected measurement point given a track trajectory.

events with a reconstructed vertex and at least two reconstructed tracks. This measurement uses data taken in 2015 at low-μ i.e. with a low number of pile-up events. The efficiency becomes higher than 99% for vertices with more than four tracks.

4.2 Leptons

Despite the physics definition of a lepton (electrons-, muons-, τ-leptons and neutrinos), in ATLAS analyses the term often refers to long lived charged leptons. In particular in this thesis, the term "lepton" includes only electrons and muons because τ-leptons decay before they can be detected and neutrinos do not leave any signal in the detector. In the next sub-section there will be only the descriptions of the reconstruction, identification and selection criteria for electrons and muon. The description of the τ-leptons will be skipped since they are not used in the analysis described in this thesis.

4.2.1 Electrons

Electrons passing through the ATLAS detector leave a track in the ID and then deposit almost all their energy in the EM calorimeter [5, 6]. For this reason the reconstruction of an electron candidates is based on three fundamental components: localised cluster of energy deposits found within the electromagnetic calorimeter, charged-particle tracks identified in the ID and close matching in $\eta \times \phi$ space of the tracks to the clusters. The tracks associated to an electron candidate need to be compatible with the primary vertex of the hard scattering and satisfy the requirements on the impact parameters, i.e. $d_0/\sigma_{d_0} < 5$ and $|z_0 \sin\theta| < 0.5$ mm, where σ_{d_0} is the d_0 uncertainty. Figure 4.2 shows a schematic illustration of the elements that enter into the reconstruction and identification of an electron.

To reconstruct the electromagnetic clusters, the $\eta \times \phi$ space of the EM calorimeter is divided into a grid of 200×250 elements, called *towers*, of size $\Delta\eta \times \Delta\phi = 0.025 \times 0.025$. The energy of the tower is the sum of the energy collected in the presampler and the other calorimeter layers. Electromagnetic-energy cluster candidates are then seeded from localised energy deposits using a *sliding-window algorithm* [7] whose summed transverse energy exceeds 2.5 GeV. The algorithm is based on the clustering of calorimeter cells within rectangles of size 3×5 tower in the $\eta - \phi$ space.[4] A removal algorithm is applied on close-by seed clusters. In addition the cluster energy is then corrected via software algorithms to account for leakages and fluctuations in the energy deposits.

[4] The cluster window is larger in ϕ than in η because the magnetic field bends charged particles in the $R - \phi$ plane, so the resulting shower spreads more in that direction.

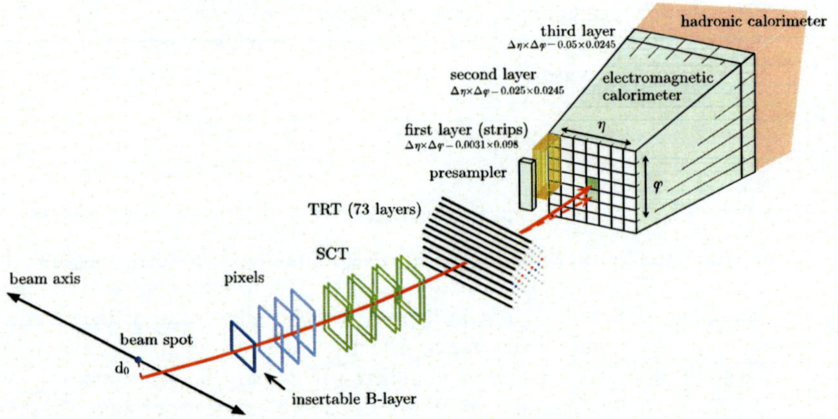

Fig. 4.2 Schematic illustration of the path of an electron through the detector. The red trajectory shows the path of an electron while the dashed red trajectory indicates the path of a photon produced by the interaction of the electron with the material of the tracking system [6]

Electron Identification

Not all the objects built by the electron reconstruction algorithm are prompt electrons. Background objects include for example hadronic jets as well as electrons from photon conversion and from semilleptonic heavy-flavour hadron decays. To reject these backgrounds, identification algorithms based on discriminating variables are applied. The baseline identification algorithm used for Run 2 data analysis is the likelihood-based (LH) method [8] based on a multivariate analysis technique. The variables used to discriminate against the different background sources describe the longitudinal and lateral shapes of the EM showers in the calorimeters, the properties of the tracks in the ID and the matching between tracks and energy clusters. Three levels of identification operating points are provided, referred to as *Loose*, *Medium* and *Tight*, in order of increasing background rejection. Each operating point uses the same variables to define the LH discriminant, but the selection on this discriminant is different for each operating point. The efficiencies of the identification criteria are determined in data and in the simulated samples. The efficiencies are calculated as the ratio of the number of electrons passing a certain identification selection (numerator) to the number of electrons with a matching track passing the track quality requirements (denominator). Figure 4.3 shows identification efficiencies as a function of the electron transverse energy E_T and pseudorapidity η considering $Z \rightarrow ee$ events with the *tag-and-probe method* [8]. The identification efficiencies for electrons with $E_T = 40$ GeV are 75% for the Tight working point and 90% for the Loose working point. Figure 4.3b shows that going from the barrel ($|\eta| < 1.2$) to the end-caps there is a drop in efficiency. The data-to-simulation ratios, the so-called scale factors, are used in physics analyses to correct the yield of electrons in simulated samples.

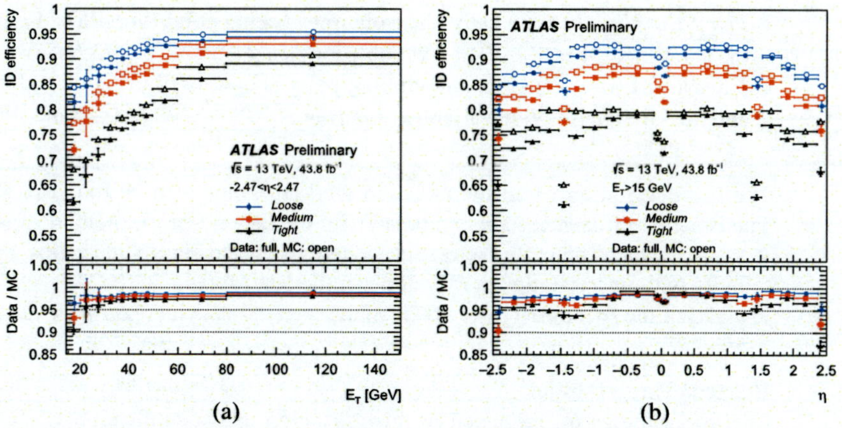

Fig. 4.3 Electron identification efficiencies in $Z \to ee$ events as a function of transverse energy E_{T}, integrated over the full pseudo-rapidity range (**a**), and as a function of pseudo-rapidity η for electrons with $E_{\mathrm{T}} > 15$ GeV (**b**). The efficiencies are shown in data and MC for three operating points, Loose (blue points), Medium (red points) and Tight (black points). For both plots, the bottom panel shows the data-to-simulation ratios [9]

Electron Isolation

Due to the high activity during a *p-p* collisions, it is challenging to distinguish prompt production of electrons from background processes such as semileptonic decays of heavy quarks, hadrons misidentified as leptons and photons converting into electron-positron pairs. The typical signature of prompt electrons, which are considered as "signal" events, is represented by little activity in the area surrounding the candidate object. The variables that quantify the amount of activity near the candidate object are evaluated by summing the transverse energies of clusters or the transverse momenta of tracks in a cone ΔR around the direction of the electron candidate, excluding the candidate itself. The main isolation variables are:

- *track-based isolation variable* $p_{\mathrm{T}}^{\mathrm{varcone20}}$: it is computed by summing the transverse momentum p_{T} of selected tracks, with a $p_{\mathrm{T}} > 1$ GeV and $|\eta| < 2.5$, within a cone centered around the electron track direction, excluding the electron track itself. The cone size ΔR shrinks for large transverse momentum of the electron and it is defined as $\Delta R = \min(10/p_{\mathrm{T}}[\mathrm{GeV}], 0.2)$;
- *calorimeter-based isolation variable* $E_T^{\mathrm{topocone20}}$: it is built by summing the transverse energy of topological clusters in a cone of size $\Delta R = 0.2$ around the electron, after subtracting the contribution from the energy deposit of the electron itself. The variable is sensitive to pile-up effects, underlying events and possible leakages. Corrections are applied to take into account these effects.

The isolation criteria are defined calculating the ratio of the track- or calorimeter-based variables to the transverse momentum of the electron ($p_{\mathrm{T}}^{\mathrm{varcone20}}/p_{\mathrm{T}}^{e}$, $E_T^{\mathrm{topocone20}}/p_{\mathrm{T}}^{e}$). In the $VH(b\bar{b})$ analysis, two electron isolation working points are applied:

- *Fixed-Cut High Pt Calo Only*: the isolation working point is optimised using only the calorimeter-based isolation variable requiring $E_T^{\text{topocone20}} < 3.5$ GeV;
- *Fixed-Cut Loose*: the isolation working point is optimised using the calorimeter- and track-based isolation variables requiring $E_T^{\text{topocone20}}/p_{\text{T}}^e < 0.20$ and $p_{\text{T}}^{\text{varcone20}}/p_{\text{T}}^e < 0.15$.

The full description of all the isolation working points can be found in Ref. [6]. The isolation efficiencies are computed with the tag-and-probe method considering $Z \rightarrow ee$ events. The scale factors, evaluated as data-to-simulation ratios, are close to unity with uncertainties mostly at per-mille level for $p_{\text{T}} > 10$ GeV, while in the low part of the spectrum ($p_{\text{T}} < 10$ GeV) the uncertainties increase with values up to 5–8%.

Electron Uncertainties
The uncertainties on the electron reconstruction and identification efficiency are evaluated in bins of transverse momentum and pseudorapidity from data and MC simulations considering $Z \rightarrow ee$, $W \rightarrow e\nu$ and $J/\psi \rightarrow ee$ events [6]. The systematic uncertainties on the reconstruction efficiency dominates at low p_{T}^e values, they are of the order of 1% for p_{T}^e = 15-30 GeV bin and of the order of per-mille level at high p_{T}^e values. The systematics uncertainties on the identification efficiency for the Loose (Tight) working point range from 3% (4%) at $p_{\text{T}}^e = 4.5$ GeV to 0.1% (0.3%) at $p_{\text{T}}^e = 40$ GeV.

The uncertainties on the electron energy scale are determined comparing the invariant mass distribution in data and MC simulation considering $Z \rightarrow ee$ events. The accuracy of the energy scale measurement ranges from 0.04% to 0.2% depending on $|\eta|$.

All the electron uncertainties are included in the fit model of the $VH(b\bar{b})$ analysis presented in this document. However, the $VH(b\bar{b})$ analysis is very weakly sensitive to the electron systematic uncertainties since they have a negligible impact on the cross-section uncertainties.

4.2.2 Muons

Muons are reconstructed and identified using the information from the ID and from the MS [6]. The reconstruction is first performed independently in the ID and MS and then the information from the subdetectors is further combined to form the muon tracks. In the ID, muons are reconstructed like any other charged particles as described in Sect. 4.1. Muon reconstruction in the MS starts with a search for hit patterns inside each muon chamber to form segments. The track candidates are then build by fitting together hits from segments in different layers. The combined track is formed using the hits from both the ID and the MS. During the global fit, MS hits may be added or removed from the track to improve the fit quality. Most muons are reconstructed following an *outside-in* pattern recognition, in which the muons are

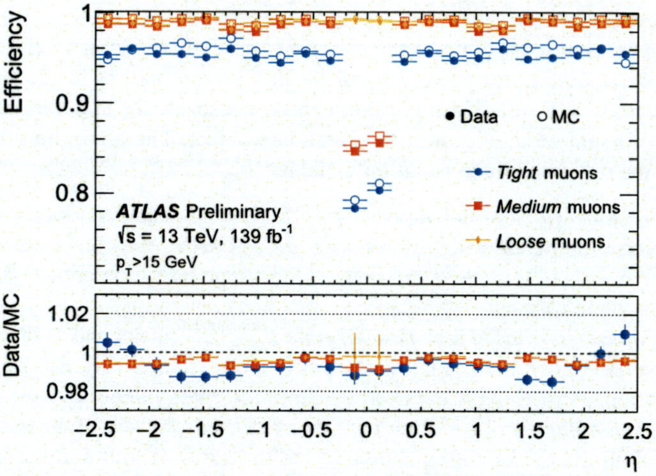

Fig. 4.4 Muon reconstruction efficiency for the Loose, Medium and Tight identification algorithms measured in $Z \to \mu\mu$ events as a function of the pseudorapidity of muon with $p_T > 15$ GeV. The prediction by the MC simulation is depicted as open circles, while filled dots indicate the observation in collision data. The bottom panel shows the ratio between expected and observation efficiencies [11]

first reconstructed in the MS and then extrapolated inward and matched to an ID track. An *inside-out* combined reconstruction is used as a complementary approach.

The muon reconstruction efficiency is measured from $Z \to \mu\mu$ and $J/\Psi \to \mu\mu$ events and it is close to 99% for $|\eta| < 2.5$ and $p_T > 5$ GeV [6].

Muon Identification

Prompt muons are separated by muons from pion and kaon decays applying identification criteria. Muons candidates originating from decays of charged hadrons in the ID can be easily recognized from the poor fit quality of the resulting combined track and from the incompatibility between the momentum measured in the ID and in the MS. To further separate muons produced in the hard scattering by the one produced in the semi-leptonic decays, especially in hadronic jets, isolation requirements and cuts on the impact parameters d_0 and z_0 are used. Differently from the electron case, the muon identification relies on a cut-based analysis. Three identification selections are provided: *Loose*, *Medium* and *Tight* with an increasing rejection power against pion and kaon decays. These categories are inclusive so muons identified with tighter requirements are also included in the looser categories. The identification efficiency of a muon in the 20 GeV $< p_T <$ 100 GeV range and $|\eta| < 2.5$ is about 98%, 96% and 92% for the Loose, Medium and Tight requirements respectively [11], as can be seen in Fig. 4.4. The measurement has been performed with a tag-and-probe method [6] on $Z \to \mu\mu$ events. There is a drop in efficiency for the Medium and the Tight working points due to the acceptance losses in the region at $\eta \sim 0$, where the MS is only partially equipped with muon chambers.

Muon Isolation

The measurement of the detector activity around the muon candidate, the so-called *muon isolation*, is a powerful tool for non prompt muons inside jets. As in case of electrons, the isolation algorithm can be constructed using variables that describe the amount of activity near the candidate object. The algorithms are mainly based on the following three isolation variables:

- *track-based isolation variable* $p_{\mathrm{T}}^{\mathrm{varcone30}}$ is defined as the scalar sum of the transverse momenta of the tracks with $p_{\mathrm{T}} > 1$ GeV in a cone of size $\Delta R = \min(10$ GeV$/p_{\mathrm{T}}^{\mu}, 0.3)$ around the muon with transverse momentum p_{T}^{μ}, excluding the muon track itself.
- *calorimeter-based isolation variable* $E_{T}^{\mathrm{topocone20}}$ is defined as the sum of the transverse energy of topological clusters in a cone of size $\Delta R = 0.2$ around the muon, after subtracting the contribution from the energy deposit of the muon itself.
- *fixed radius track isolation* $p_{\mathrm{T}}^{\mathrm{cone20}}$ is defined as the sum of the transverse momenta of all the tracks within a cone of radius $\Delta R = 0.2$ around the muon track, after subtracting the muon contribution itself.

As in case of the electrons, the isolation criteria are determined using the relative isolation variables defined as the ratio of the track- or calorimeter-based isolation variables to the transverse momentum of the muon ($p_{\mathrm{T}}^{\mathrm{varcone30}}/p_{\mathrm{T}}^{\mu}$, $E_{T}^{\mathrm{topocone20}}/p_{\mathrm{T}}^{\mu}$, $p_{\mathrm{T}}^{\mathrm{cone20}}/p_{\mathrm{T}}^{\mu}$). In the $VH(b\bar{b})$ analysis, the applied isolation working points are:

- *Fixed-Cut High Pt Track Only isolation*: the isolation working point is optimised using only the track information requiring $p_{\mathrm{T}}^{\mathrm{cone20}} < 1.25$ GeV;
- *Fixed-Cut Loose isolation*: the isolation working point is optimised using the track and the calorimeter information requiring $E_{T}^{\mathrm{topocone20}}/p_{\mathrm{T}}^{\mu} < 0.3$ and $p_{\mathrm{T}}^{\mathrm{varcone30}}/p_{\mathrm{T}}^{\mu} < 0.15$.

The efficiencies for the different working points are measured in data and simulation in $Z \to \mu\mu$ events using the tag-and-probe method [6]. Scale factors have been calculated as data-to-simulation ratios and they have been derived to correct the muon isolation efficiency in simulation to match the measured data for each working point. The scale factors are compatible within the uncertainties, typically 0.5-2%.

Muon Uncertainties

The muon momentum scale and resolution uncertainties are evaluated considering $J/\psi \to \mu\mu$ events in the $p_{\mathrm{T}}^{\mu} < 20$ GeV region and $Z \to \mu\mu$ events in the $p_{\mathrm{T}}^{\mu} \geq 20$ GeV region [6]. The systematic uncertainties on the momentum scale varies from 0.05% in the barrel region ($|\eta| < 1$) to 0.1% (0.3%) in the region $|\eta| \sim 2.5$ for $Z \to \mu\mu$ ($J/\psi \to \mu\mu$) decays. The momentum resolution in the central region is measured to be 1.7 % (2.3%) for muons from $J/\psi \to \mu\mu$ ($Z \to \mu\mu$) decays, while in the region $|\eta| > 2.2$, the p_{T} resolution for muons from $Z \to \mu\mu$ decays is 2.9%.

The systematic uncertainties on the identification efficiency is at per-mille level over a wide range of p_{T} and reaches the percent level in the high-p_{T} region.

4.3 Jets

QCD partons manifest themselves in the detector as a direction spray of tracks and calorimeter deposits known as jets. Jets are typically reconstructed by grouping energy deposits in the calorimeters creating the so-called *topological-clusters* or *topo-clusters* (Sect. 4.3.1). For some measurements the calorimeter jets are not optimal and the information is complemented exploiting track-jets (Sect. 4.3.7). Track-jets are built using the same algorithms applied for calorimeter jets but using tracks emerging from the hard-scattering vertex instead of using energy deposits. Clusters are the inputs of the jet reconstruction algorithms. In the analysis presented in this document, two different jet reconstruction algorithms have been used: the k_t and the anti-k_t algorithms (Sect. 4.3.2).

4.3.1 Topological Cluster Formation and Calibration

Crossing the detector, interacting particles deposit their energy in several cells of the calorimeters. The *topological-clustering algorithm* [12] is used to group the associated cells into *cluster* following the spatial signal-significance patterns generated by particle showers. The observable controlling this cluster formation is the *cell signal significance* $\zeta^{\mathrm{EM}}_{\mathrm{cell}}$ which is defined as the ratio of the cell signal to the average (expected) noise is this cell.[5] The clustering begins identifying a cell having a cell signal significance grater than four and adding to this seed cell the neighbouring cells. If the neighbouring cell significance exceeds two, the procedure is iteratively applied to further neighbours. Each seed cell and its neighbouring cells form a *proto-cluster*. Two pro-clusters are merged if a direct neighbour is a seed cell itself or if two seed cells share a direct neighbouring cell with significance exceeding two. The algorithm stops when no new neighbouring cells are added to the proto-cluster. This algorithm has the advantage that tails of the showers are not discarded but, at the same time, large clusters including showers from different particles are built. This disadvantage is remedied by splitting the proto-clusters with two or more local maxima. The local maxima are considered as new seeds and the clustering is done considering only cells inside the proto-cluster. The obtained proto-clusters are the *topo-clusters* used as inputs for the object reconstruction, i.e. jet reconstruction algorithm.

The energy of the topo-clusters is the sum of the energy of the single cells. Both the cell signal and the average cell noise are measured at the electromagnetic energy scale reconstructing the energy deposited by electrons and photons. Thus the topo-clusters energies must be corrected for the lower response of the calorimeters to hadrons. The *local hadronic cell weighting* (LCW) calibration [12] is the strategy used by the ATLAS Collaboration. Topo-clusters are classified as electromagnetic

[5] $\zeta^{\mathrm{EM}}_{\mathrm{cell}} = \frac{E^{\mathrm{EM}}_{\mathrm{cell}}}{\sigma^{\mathrm{EM}}_{\mathrm{noise,cell}}}$ where $E^{\mathrm{EM}}_{\mathrm{cell}}$ is the cell signal and $\sigma^{\mathrm{EM}}_{\mathrm{noise,cell}}$ is the average expected cell noise. Both cell signal and noise are measured on the electromagnetic energy scale.

or hadronic based on the energy shower observed in the calorimeters. The cluster energy is then corrected by a moment-dependent factor derived using simulations of single pions and tested in situ using test-beam data. These corrections account for the non-compensating calorimeter response, the signal losses due to noise threshold effects inherent to the clustering algorithm and the signal losses due to the energy lost in inactive material.

4.3.2 Jet Reconstruction Algorithms

Calibrated topo-clusters are used as input objects for the jet reconstruction algorithm. In the ATLAS Collaboration the clustering algorithm mainly used is the anti-k_t algorithm [13]. The algorithm lies on the definition of distance d_{ij} between two objects i and j and on the distance d_{iB} between object i and the beam B as:

$$d_{ij} = \min\left(k_{ti}^{2p}, k_{tj}^{2p}\right) \cdot \Delta R_{ij}^2 \tag{4.1a}$$

$$d_{iB} = k_{ti}^{2p} \cdot R_i^2 \tag{4.1b}$$

where ΔR_{ij} is the angular distance[6] between two objects, $k_{ti/j}$ their transverse momenta, R_i the *radius parameter* of object i and p governs the relative power of the energy versus the geometrical scales. The jet definition of the anti-k_t algorithm is obtained for $p = -1$. The case $p = 0$ corresponds to the inclusive Cambridge/Aachen algorithm, while $p = 1$ corresponds to the k_t algorithm [14]. For each object i, the clustering proceeds as follows:

1. compute d_{iB} and all d_{ij};
2. identify the smallest of the distances;
   - if the smallest distance is d_{ij}, merge i with j into a single new object;
   - if the smallest distance is d_{iB}, consider i to be a jet and remove it from the list of objects;
3. the algorithm restarts from point 1 with the updated list of input objects until no object is left in the list.

Once this clustering procedure is done, the jet four-momentum is obtained as the sum of the cluster four momenta.

In the $VH(b\bar{b})$ analysis, the anti-k_t algorithm is used to reconstruct both calorimeter and track-jets collection.[7] Additionally, two kinds of calorimeter jets are reconstructed with the anti-k_t algorithm: the small-radius jets (small-R jets) reconstructed

[6] The angular distance ΔR_{ij} between two objects i and j is defined as the angular distance between the two energy barycentres $\Delta R_{ij} = \sqrt{(\eta_i - \eta_j)^2 + (\phi_i - \phi_j)^2}$.

[7] In case of track-jets collections, tracks are used as input of the jet reconstruction algorithm.

with a radius parameter of $R_i = R = 0.4$ and large-radius jets (large-R jets) with $R_i = R = 1.0$. Differently, the k_t algorithm is applied to reject the jet component belonging to pile-up. The key property of the anti-k_t algorithm is that the jet shape is resilient with respect to the soft radiation but, at the same time, it is flexible with respect to the hard radiation. Oppositely, the k_t algorithm clusters energy deposits into jets starting from thc softest part up to the hardest one.

4.3.3 Small-R Jets Calibration

Once jets are reconstructed using the anti-k_t algorithm, a calibration procedure is employed aiming to restore the jets energy scale (JES) of the small-R jets to that of jets reconstructed from simulated stable particles. The JES calibration [15] proceeds in six steps (Fig. 4.5) and each stage of the calibration corrects the full four-momentum scaling the jet p_{T}, energy and mass.

The first correction is the *origin correction* which recalculates the four-momentum of the jets to point to the hard-scattering primary vertex rather than the center of the detector. This correction does not affect the jet energy but it improves the η resolution of jets. It is then followed by two pile-up corrections: *jet area-based pile-up correction* [16] and the *residual pile-up correction*. The former method subtracts the per-event pile-up contribution, defined as the expected energy density of soft contributions multiplied by the jet area, to the p_{T} of each jet. The latter aims at subtracting the residual pile-up dependence. This correction depends on the exact number of primary vertices (N_{PV}) and the average number of interaction per bunch crossing. Then, the *absolute MC-based calibration* corrects the jet four-momentum to the particle-level energy scale and it accounts for biases in the jet η reconstruction. Such biases are due to the transitions between different calorimeter technologies and sudden changes in the calorimeter granularity. Further improvements to the reconstructed energy and related uncertainties are achieved through calorimeter, MS and track-based variables in the *global sequential calibration*. Finally, a *residual*

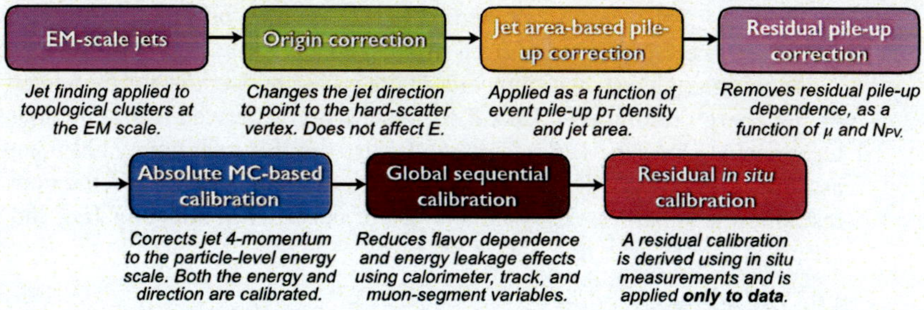

Fig. 4.5 Steps of the JES calibration procedure applied to small-R jets. Each stage of the calibration is applied to the four-momentum of the jet [15]

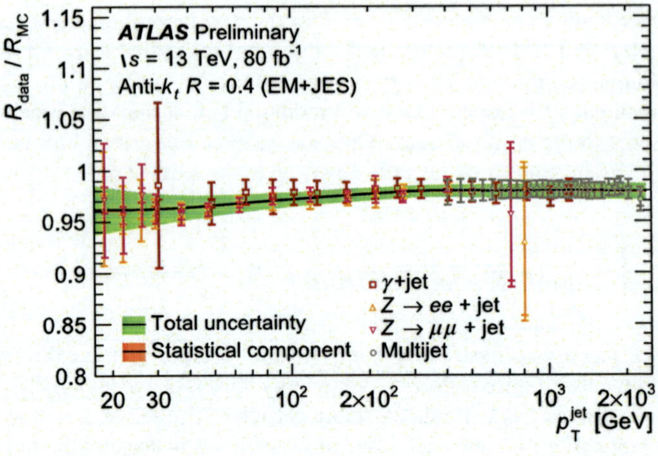

Fig. 4.6 Data-to-simulation ratio of the average jet p_T response as a function of the jet p_T. The result is based on three in-situ techniques: Z+jets (triangles), γ+jets (squares) and multi-jet balance method (circles) [17]

in-situ calibration is applied to correct jets for potential differences between data and simulation. The in-situ corrections are extracted from analyses which study γ/Z+jets or multi-jet processes, respectively. The corrections are applied only to data. An example of the jet energy scale factors extracted from in-situ calibration is shown in Fig. 4.6.

The systematic uncertainties associated to the jet energy calibration are mainly coming from the in-situ calibrations extracted from analyses that examine γ/Z+jets or multi-jet processes. These uncertainties are derived from the comparison of data with MC simulations by varying different parameters as the event selection, the MC generator or the identification criteria in the analysis. The uncertainty contribution is evaluated separately for each in-situ method. Another source of uncertainty related to the jet energy comes from the flavor dependence that accounts for differences in the jet response and simulated jet composition of light-quark, b-quark, and gluon-initiated jets. The remaining uncertainties comes from pile-up or η-intercalibration. All these uncertainties are referred to as jet energy scale (JES) uncertainties.

An additional systematic uncertainty is added to account for the uncertainty on the jet energy resolution (JER). This uncertainty is derived using the *dijet balance* technique [18]. As the JES uncertainties, also the JER uncertainty has many sources as, for example, the difference between data and simulation when the nominal data resolution is superior. The main sources of uncertainty affecting JES and JER are shown in Fig. 4.7 as a function of jet-p_T.

Pile-Up Jet Suppression

Before using the calibrated small-R jets in the analysis, a pile-up suppression algorithm is applied. To suppress jets from pile-up, a new likelihood-based discriminant

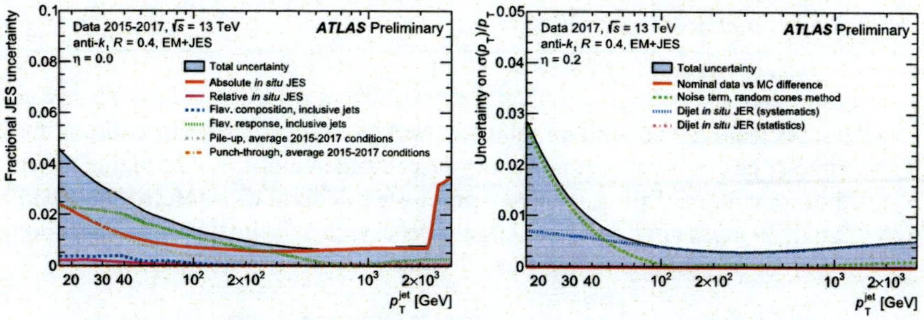

(a) (b)

Fig. 4.7 **a** Fractional jet energy scale systematic uncertainty components as a function of jet-p_T for small-R jets at $\eta = 0.0$ and, **b** jet energy resolution uncertainty as a function of jet-p_T for small-R jets at $\eta = 0.2$. The total uncertainty (all components summed in quadrature) is shown as a filled region topped by a solid black line [17, 19]

called Jet Vertex Tagger (JVT) [20] is used in Run 2. The tool uses information about the primary vertex, jet and track p_T. The JVT values range from 0 to 1, where jets with values close to 1 are signal jet candidate while jets close to 0 are from pile-up. The default JVT cut is at 0.59 with an efficiency for selected hard-scattering jets of 92% and a fake rate of about 1%. In the analysis presented in this document, the JVT tool is only applied to small-R jet with $p_T < 120$ GeV and $|\eta| < 2.5$.

4.3.4 Large-R Jets

The high centre-of-mass energy of the LHC has opened new kinematic regimes to experimental study. The new phase space available for the production of SM particles with significant Lorentz boosts necessitates new techniques. When sufficiently boosted, the decay products of the boosted particle, i.e. the Higgs boson, can become collimated to the point that standard reconstruction techniques begin to fail. The angular separation of the decay products b_1 and b_2 is approximately:

$$\Delta R(b_1, b_2) \approx \frac{2m}{p_T} \tag{4.2}$$

where m and p_T are the mass and the transverse momentum of the decaying particle. From this rule, for the Higgs boson candidate with a $p_T > 250$ GeV, a large-R jet with radius parameter of 1 is needed.

4.3.5 Large-R Jet Mass

The mass of the large-R jets is a powerful tool for distinguishing massive particle decays from the continuum multi-jet background. A priori the invariant mass of the large-R jet can be determined using the calorimeter information taking the energy and the momentum of all associated topo-clusters. This mass is referred to as *calorimeter-based jet mass* m^{calo} and for a large-R jet J with calorimeter-cell cluster constituents i with energy E_i and momentum $\vec{p}_i$ ($|\vec{p}_i| = E_i$) it is defined as:

$$m^{\text{calo}} = \sqrt{\left(\sum_{i \in J} E_i\right)^2 - \left(\sum_{i \in J} \vec{p}_i\right)^2} \tag{4.3}$$

However, the resolution degrades at high jet p_{T} because the angular separation of the decay products of the boosted particle falls below the calorimeter granularity (Fig. 4.8b). For this reason, the tracking information provides an excellent directional resolution and the *track-assisted jet mass* m^{TA} is defined as:

$$m^{\text{TA}} = \frac{p_{\text{T}}^{\text{calo}}}{p_{\text{T}}^{\text{track}}} \times m^{\text{track}} \tag{4.4}$$

where $p_{\text{T}}^{\text{calo}}$ is the transverse momentum of the large-R jet, $p_{\text{T}}^{\text{track}}$ and m^{track} are the transverse momentum and mass of the four-vector sum of tracks associated to the large-R jet. In this mass definition, the ratio of the transverse momenta is used to correct for charged-to-neutral fluctuations. Figure 4.8a shows the comparison of m^{calo}, m^{TA} and m^{track} before and after any calibration in a very high p_{T} range (1.6 TeV $< p_{\text{T}} <$ 1.8 TeV). The peak position and the width of the track-assisted jet mass are comparable to the calorimeter-based jet mass and significantly better than the track-only mass. However, the resolution of m^{TA} becomes worse than the one of the m^{calo} for low jet p_{T} (Fig. 4.8b). The solution for this issue is to combine the calorimeter-based jet mass with the track-assisted jet mass. Thus the *combined jet mass* m^{comb} is defined as:

$$m^{\text{comb}} = a \cdot m^{\text{calo}} + b \cdot m^{\text{TA}} \tag{4.5}$$

with $a = \sigma_{\text{calo}}^{-2}/(\sigma_{\text{calo}}^{-2} + \sigma_{\text{TA}}^{-2})$ and $b = \sigma_{\text{TA}}^{-2}/(\sigma_{\text{calo}}^{-2} + \sigma_{\text{TA}}^{-2})$, where σ_{calo} (σ_{TA}) is the resolution function of m^{calo} (m^{TA}). The combined jet mass smoothly interpolates between $m^{\text{comb}} \sim m^{\text{calo}}$ at low p_{T} and $m^{\text{comb}} \sim m^{\text{TA}}$ at high p_{T}.

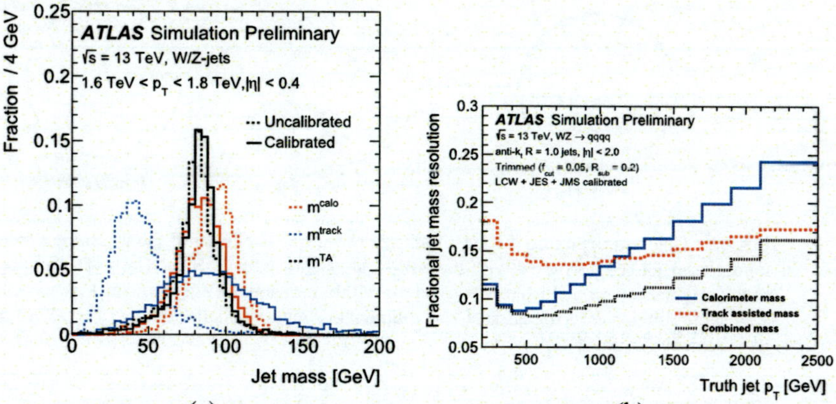

(a) (b)

Fig. 4.8 **a** Uncalibrated (dashed line) and calibrated (solid line) reconstructed jet mass distribution for the different large-R jet mass definition [21]. **b** Fractional jet mass resolution vs. the truth jet mass transverse momentum for the three large-R jet mass definition [22]

4.3.6 Large-R Jets Grooming and Calibration

The large center-of-mass energy at LHC enables the production of Lorentz-boosted heavy particles, whose decay products can be reconstructed as one large-R jet. At the same time, multiple p-p interactions per bunch crossing produce soft particles unrelated to the hard scattering that contaminate jets. To remove the contribution of soft and pile-up particles, *grooming techniques* are applied [23]. In the analysis presented in this document, the grooming technique applied is the *trimming technique* [24]. This algorithm is implemented by reclustering the constituents of the initial jets using the k_t algorithm into small-R jets with a radius parameter R_{sub}. Then, any subjets that has a $p_{\mathrm{T}}^{i}/p_{\mathrm{T}}^{\mathrm{jet}}$ less than f_{cut} are removed, where p_{T}^{i} and $p_{\mathrm{T}}^{\mathrm{jet}}$ are the transverse momentum of the ith subjet i and of the parent jet, and f_{cut} is a parameter of the method which is typically a few percent. The remaining constituents form the trimmed jet. This procedure is illustrated in Fig. 4.9. The trimming algorithm applied in the $VH(b\bar{b})$ analysis has $R_{sub} = 0.2$ and $f_{cut} = 5\%$.

The effect of the trimming technique can be observed in Fig. 4.10. The plots show the anti-k_t $R = 1.0$ leading jet mass distribution before (left) and after (right) jet grooming in $Z' \rightarrow t\bar{t}$ events for different mean number of interactions per bunch crossing μ. Applying the trimming technique, the effect of pile-up is highly reduced and, the top mass peak and the smaller peak at around the W mass is now visible.

The JES and the jet mass scale (JMS) are corrected to the particle level scale using matched truth jet collections. All the corrections are applied as a function of the reconstructed jet energy, mass and pseudorapidity. Finally the jets are calibrated *in-situ* using response measurement in data. The *in-situ* correction evaluated as data-to-simulation ratio is applied only to data for taking into account the residual

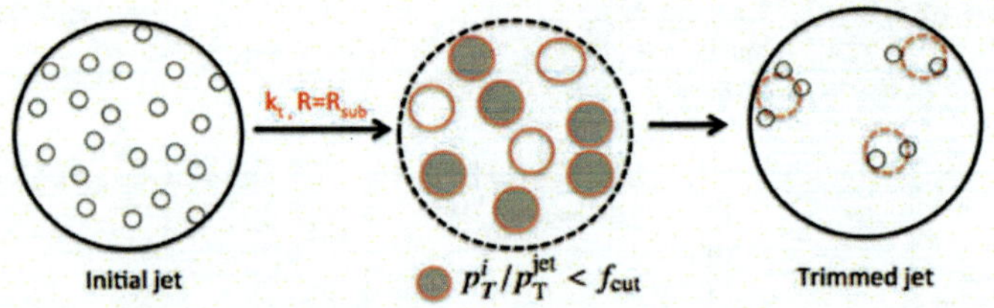

Fig. 4.9 Diagram depicting the jet trimming procedure. The input of the trimming procedure is a topo-cluster jet reconstructed with the anti-k_t algorithm with $R = 1.0$ (left). The jet constituents of the original jet are reclustered using k_t algorithm into smaller subjets with $R_{sub} = 0.2$ (center). Any subjets that is carrying a transverse momentum fraction smaller than $f_{cut} = 5\%$ of the parent jet is removed (right) [23]

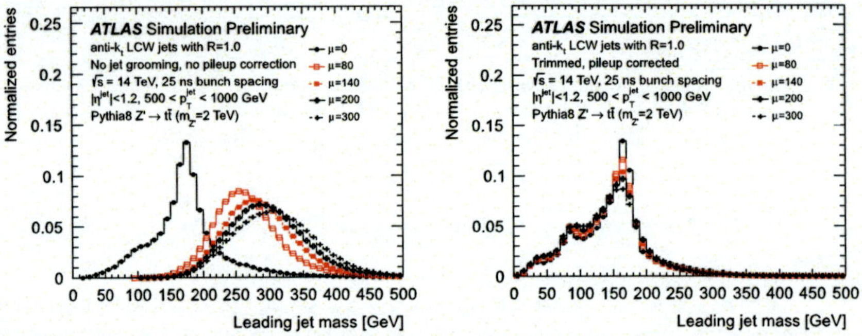

(a) (b)

Fig. 4.10 Mass distribution of anti-k_t large-R jet with (right) and without (left) applying the trimming technique [25]

mismodelling in the response. The correction factor relies on the transverse momentum balance in a variety of final states: Z +jets, γ+jets and multi-jets. The absolute calibration is derived from a statistical combination of these three measurements. Additionally, a relative intercalibration, derived using dijets events, propagates the well-measured central JES into the forward region of the detector. The JES correction is applied as four-momentum scale factor to jets in data and it also affects the jet mass. The calibration factor is defined as the double ratio between data and MC and it is applied only to data. Figure 4.11 summarizes the large-R jet calibration chain.

The uncertainties related to large-R jets can be splitted into *scale uncertainties*, which include uncertainties on the JES and on the JMS, and *resolution uncertainties*, which take into account the uncertainty on the jet mass and energy (p_{T}) resolution.

The JES and JMS uncertainties are derived *in-situ* using the R_{trk} *method* [23] which relies on the assumption that the uncertainties on the tracking and calorimeter measurements are largely uncorrelated. The basic element of the R_{trk} method is the ratio between the calorimeter response and the tracker response, i.e. for the jet energy scale uncertainty the ratio is defined as:

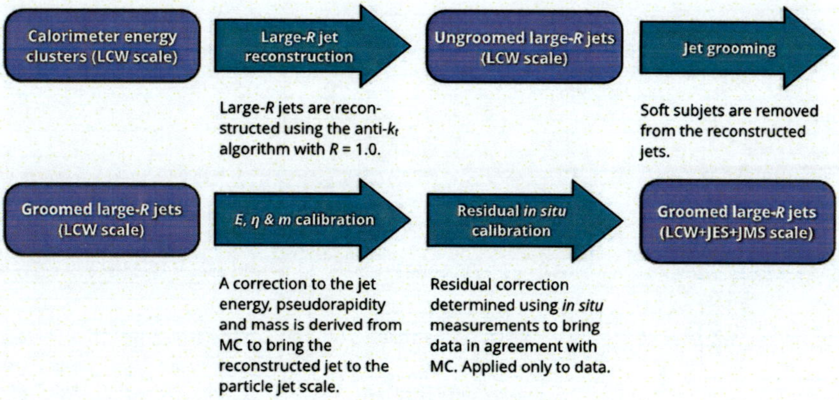

Fig. 4.11 Overview of the large-R jet calibration procedure [26]

$$r_{\text{trk}}^{p_{\text{T}}} = \frac{p_{\text{T}}^{\text{calo}}}{p_{\text{T}}^{\text{tracker}}} \tag{4.6}$$

where $p_{\text{T}}^{\text{calo}}$ is the large-R transverse momentum computed with the calorimeter information and $p_{\text{T}}^{\text{tracker}}$ is the transverse momentum computed with tracks. The $r_{\text{trk}}^{p_{\text{T}}}$ variable is then used to evaluate the double ratio:

$$R_{\text{trk}}^{p_{\text{T}}} = \frac{r_{\text{trk}}^{p_{\text{T}},\text{data}}}{r_{\text{trk}}^{p_{\text{T}},\text{MC}}} \tag{4.7}$$

where $r_{\text{trk}}^{p_{\text{T}},\text{data}}$ and $r_{\text{trk}}^{p_{\text{T}},\text{MC}}$ are the ratios computed with data and with simulated samples, respectively. The double ratio $R_{\text{trk}}^{p_{\text{T}}}$ should be equal to the unity if the detector is well modelled. This double ratio is also evaluated comparing the nominal simulation with the alternative simulations to take into account uncertainties arising from differences in the fragmentation and hadronisation models. The systematic uncertainties are computed in bins of the jet mass over transverse momentum m/p_{T}. This method used for the scale uncertainty is versatile and it is applied to evaluate the JMS uncertainty.[8] Figure 4.12 shows $r_{\text{trk}}^{p_{\text{T}}}$ and r_{trk}^{m} as a function of the large-R jet p_{T} in dijets events for data and several simulations. In both cases the calorimeter-to-track ratio is grater than the unity because the information of the tracker accounts only

[8] For the computation of the JMS uncertainty, the ratio between the calorimeter and the tracker response r_{trk}^{m} is defined as:

$$r_{\text{trk}}^{m} = \frac{m^{\text{calo}}}{m^{\text{trk}}} \tag{4.8}$$

where m^{calo} is the large-R jet mass computed with the calorimeter information and m^{trk} is the large-R jet mass computed with tracks.

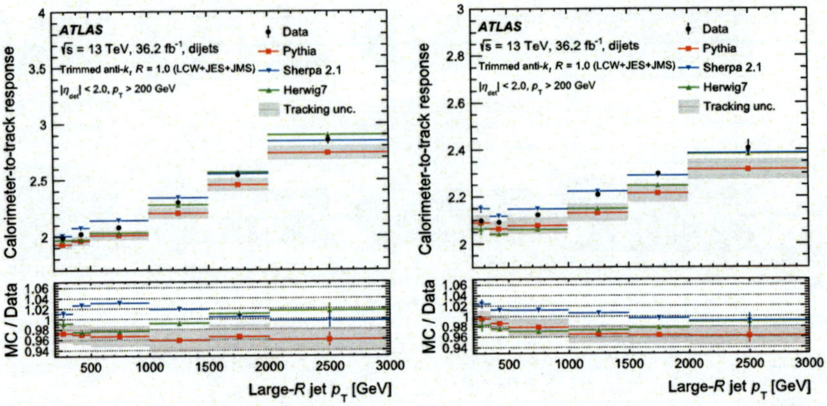

(a) (b)

Fig. 4.12 Measurement of $r_{\mathrm{trk}}^{p_{\mathrm{T}}}$ (**a**) and r_{trk}^{m} (**b**) as a function of the large-R jet p_{T} for large-R jets with $m/p_{\mathrm{T}} = 0.2$. Data are compared with three generators and with three tracking variations for the default generator PYTHIA 8. The double ratio of $r_{\mathrm{trk}}^{p_{\mathrm{T}}}$ and r_{trk}^{m} measured in simulations and data is shown in the lower panel [26]

for charged particles. The lower panel of the plots shows the double ratio between simulations and data $(1/R_{\mathrm{trk}}^{m/p_{\mathrm{T}}})$.

Using the R_{trk} method, four sources of scale uncertainty are computed:

- the *baseline variation* accounts for the deviations from unity of R_{trk} between data and the nominal simulation. This variation accounts for all the differences between data and simulated jets and the simulation of the interaction of the particle with the detector.
- the *modelling variation* accounts for the deviations of R_{trk} between the nominal and the alternative simulations. The maximum difference is taken as modelling uncertainty.
- the *tracking* uncertainty is the quadratic sum of the effect of the various tracking uncertainty sources.
- the *statistical* uncertainty accounts for the statistical uncertainty of the data and the simulated samples. Typically this uncertainty is small at low large-R jets p_{T} and relevant at very high p_{T}.

The total JES and JMS uncertainties as a function of the large-R jet p_{T} are shown in Fig. 4.13. In both cases the statistical uncertainty has a relevant effect only at high p_{T} ($p_{\mathrm{T}} > 2$ TeV). For the JES (JMS), tracking uncertainty contribution is constant for $p_{\mathrm{T}} > 400$ GeV ($p_{\mathrm{T}} > 800$ GeV) and its contribution is about 2%. The modelling uncertainty is also of the order of 2% with a larger impact at about 1 TeV. The baseline uncertainty is the main source of uncertainty and it is about 5–6%.

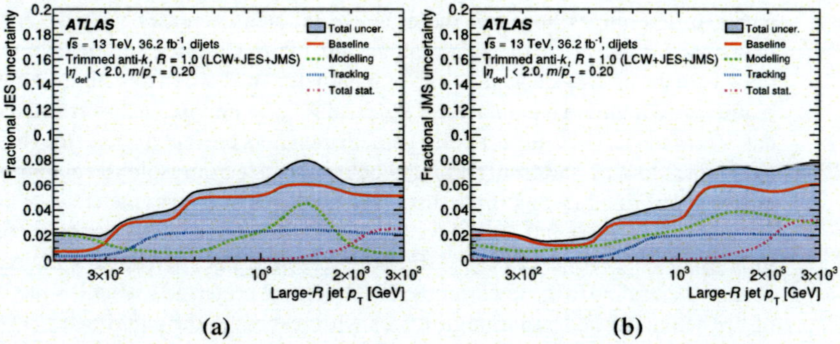

Fig. 4.13 Total uncertainty in the relative JES (**a**) and JMS (**b**) in data and simulations as a function of the large-R transverse momentum [26]

For the p_T resolution uncertainty, a gaussian with an absolute 2% uncertainty is used to smear the simulated distribution. The difference between the smeared and unsmeared distributions is taken as resolution uncertainty.

For the jet mass resolution (JMR) uncertainty, a relative jet mass resolution smearing of 20% is applied to each jets. To apply a relative smearing, resolution maps are needed. These resolution maps are computed for different jets: top-jets, W/Z-jets and Higgs-jets.

4.3.7 Track-Jets

Track-jets are jets reconstructed by applying the jet clustering algorithm on the tracks emerging from the hard-scattering vertex. The tracks used as input to the jet reconstruction algorithm are required to have:

- $p_\mathrm{T} > 0.5$ GeV and $|\eta| < 2.5$;
- at least 7 hits in the pixel and SCT detectors;
- not more than one hit in the pixel detector shared by multiple tracks;
- not more than one missing hit in the pixel detector when it is expected;
- not more than two missing hits in the SCT detector;
- the longitudinal impact parameter z_0 of the tracks is required to be $|z_0 \cdot \sin\theta| < 3$ mm.

The selected tracks are used as input of the anti-k_t algorithm. The track-jets used in this analysis are the *variable-R* (VR) *track-jets* [27] with a radius parameter defined as:

$$R_i = R(p_{\mathrm{T}i}) = \frac{\rho}{p_{\mathrm{T}i}} \tag{4.9}$$

where ρ determines how fast the effective jet size decreases with the transverse momentum of the jet. In the anti-k_t algorithm, the R_i parameter of Eq. 4.1b is substituted with the new p_T dependent radius parameter. In addition to ρ, the VR algorithm requires two additional parameters, R_{min} and R_{max} to impose the lower and the upper cut-offs of the jet size, respectively. These parameters prevent the jets from becoming too large at low p_T and from shrinking below the detector resolution at high p_T. The analysis described in this thesis uses VR track-jets with parameters $\rho = 30$ GeV, $R_{max} = 0.4$, $R_{min} = 0.02$ that have been optimized for double b-tagging efficiency in boosted $H \rightarrow b\bar{b}$ decays [28]. Figure 4.15a (red line) shows how the cone size changes according to the track-jet p_T. Track-jets are preferred to small-R jets because of the better spatial resolution of tracks with respect to the calorimeter clusters. In the $VH(b\bar{b})$ analysis they are used only to tag jets containing b-hadrons within the large-R jets since in boosted topologies there are several benefits of using track-jets for finding b-hadrons [29, 30]. Since b-tagging algorithm only uses information from the inner detector, track-jets can be decoupled from calorimeter jet finding in order to identify b-hadrons. The idea is to optimize the track-jet algorithm for b-tagging and, independently, to optimize the calorimeter jet algorithm to interpret the final state. Additionally, the track-jets are explicitly chosen to originate from the primary vertex, reducing significantly the performance dependence on pile-up. This is important when reconstructing low p_T b-hadrons produced in the opposite direction to the boost direction of the decaying particle. These low p_T b-hadrons might be lost due to the p_T-threshold imposed on calorimeters jets in order to reduce pile-up. Finally, track-jets have a good angular resolution even in environments with dense hadronic activity. Furthermore, the track-jets can easily take advantage of the small radius parameters, which lead to the best resolution and efficiency at high boost values.

Large-*R* to Track-Jets Association

A dense hadronic environment, there can be some ambiguity when trying to match the track-jets to the large-R jet. To solve this ambiguity, the *ghost association* technique can be applied to determine the jet area [31]. In this algorithm, the track-jet is added to the large-R jet as an infinite soft particle, the so-called "ghost", retaining only its spatial information. The large-R jet is then re-clustered using the anti-k_t algorithm considering as input both the calorimeter clusters and the ghost track-jets. A track-jet is considered ghost-associated with the large-R jet if its ghost version is contained in the large-R jet after reclustering. The resilience of the jet reconstruction algorithm to soft particle and the infinitesimal p_T of the ghosts ensure that no difference is obtained in the final large-R jet with respect to the one clustered without ghosts.

Flavour Tagging

The VR track-jets are used as input to the dedicated flavour tagging algorithms with the aims of correctly identifying the ones originating from an initial b-hadron decay. The characteristics of b-jets are connected to the properties of b-quarks such as their relatively high mass ($\sim$ 5 GeV), long lifetime ($\sim$ 1.5 ps) and high decay multiplicity (on average $\sim$ 5 charged particles). This leads to topologies with at least one *secondary vertex* separated from the *primary vertex* (Fig. 4.14). All these

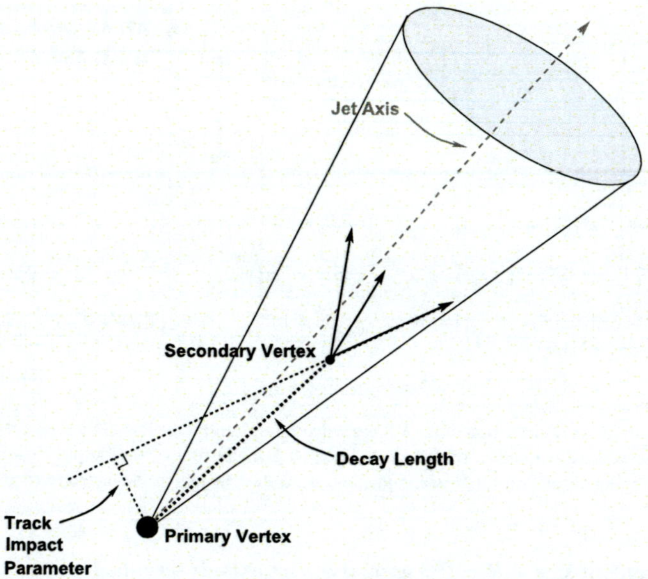

Fig. 4.14 Illustration of a jet with a secondary vertex reconstructed from displaced tracks with a large impact parameter and with a significant decay length indicating the decay of a heavy long-lived particle, i.e. a b-hadron [32]

characteristics are used by the flavour tagging algorithms to distinguish b-jets from jets originated by light flavour hadrons.

The tracks used in the flavour tagging algorithm are associated to the track-jet but they are not necessarily the same ones that compose it. The matching between the track and the track-jet is based on the angular separation ΔR considering tracks with a $p_\mathrm{T} > 1$ GeV. The ΔR selection decreases with increasing jet p_T according to Equation 4.10:

$$\Delta R(\text{track}, \text{jet}) < 0.239 + e^{-1.220-1.64\cdot 10^{-5}\cdot p_T} \; p_T[\text{MeV}] \qquad (4.10)$$

For high jet p_T, the ΔR has a plateau at a value of 0.239 (Fig. 4.15a, green line)

The b-tagging algorithm used in the $VH(b\bar{b})$ analysis is the MV2 algorithm [34]. It consists of a boosted decision tree (BDT) that combines information on the impact-parameter, on the secondary vertex and on the b-to-c-hadron decay chain. The kinematic properties of the jets (p_T and $|\eta|$) are also included in the training. The MV2 tagger has been trained on small-R calo jets with $R = 0.4$ from the hybrid $t\bar{t}$ and Z' samples where jets with $p_\mathrm{T} < 250$ GeV come exclusively from $t\bar{t}$ events and jets with $p_\mathrm{T} > 250$ GeV are taken from $Z' \rightarrow$ dijets events. In the simulation, the jet flavour labels are attributed looking at the truth particles. Jets are labelled as b-jets if at least one weakly decaying b-hadron with $p_\mathrm{T} > 5$ GeV is found within

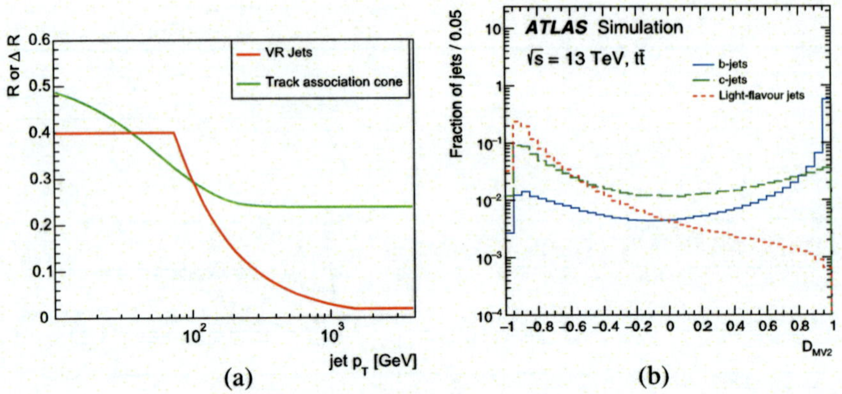

Fig. 4.15 **a** Distribution of the VR track-jet radius parameter R (red line) and of the track association cone ΔR as a function of the jet p_{T} [32]. **b** Distribution of the output discriminant of the MV2 b-tagging algorithm for b-jets, c-jets and light-flavour jets in the baseline $t\bar{t}$ simulated events [33]

a cone of size $\Delta R = 0.3$ around the jet axis. If no b-hadrons are found, c-hadrons are searched for, and afterwards τ-leptons. The jets matched with c-hadrons and τ-leptons are labelled as c-jets and τ-jets, respectively. The remaining jets are labelled as light-flavour jets. For the training , the c-jet fraction in the background sample is set to 7%, the remaining part of the background (93%) is composed by light-flavour jets. The list of input variables has been optimised to provide the best separation power between the signal and the background. The output discriminant of the MV2 algorithm $\mathrm{D}_{\mathrm{MV2}}$ for b-jets, c-jets and light-flavour jets evaluated with the baseline $t\bar{t}$ simulated events are shown in Fig. 4.15a.

A jet is defined as b-tagged if the MV2 output score exceeds a certain threshold. This threshold is defined a priori as the cut that gives a pre-determined efficiency value for b-jets when applied to a $t\bar{t}$ sample. The $VH(b\bar{b})$ analysis uses a MV2 cut that corresponds to an average 70% b-tagging efficiency [33]. The rejection rate, defined as the inverse mistag efficiency, for c-jets and light-flavour jets has been determined to be 9 and 304, respectively. Both the tagging and the mistagging efficiencies are furthermore measured on data to correct the simulation by applying data-to-simulation scale factors binned in jet p_{T}. Although the MV2 tagger has been trained on small-R calo jets and used for VR track-jets, the calibrations factors have been computed using VR track-jets. The uncertainty on these factors introduces a systematic uncertainty in the analysis [35].

4.4 Missing Transverse Momentum

Particles emerging from the hard-scattering interaction that are solely weakly interacting, such as neutrinos, escape the direct detection. Their presence can be only inferred from the missing transverse momentum $E_{\mathrm{T}}^{\mathrm{miss}}$ which is defined as an imbalance in the plane transverse to the beam line. Following the conservation of momentum, the total transverse momentum of all particles in the event should be zero, so the total momentum associated with these non-interacting particles can be computed as the negative vector of the transverse momenta of the other particles. The $E_{\mathrm{T}}^{\mathrm{miss}}$ is computed combining the information from the calorimeter and the muon spectrometer for all the objects (leptons, jets associated to the primary vertex, etc) [36]. A term is added to account for good quality soft tracks not associated to any visible particle produced in the hard processes. The $E_{\mathrm{T}}^{\mathrm{miss}}$ is defined as:

$$E_T^{miss} = -\sum_{muons} p_T^{\mu} - \sum_{electrons} p_T^{e} - \sum_{photons} p_T^{\gamma} - \sum_{taus} p_T^{\tau} - \sum_{jets} p_T^{jets} - \sum_{soft\,trks} p_T^{trk} \tag{4.11}$$

Energy deposits in the calorimeters and tracks are matched to reconstructed objects in an order chosen to minimise double-counting of elements. The order used for $E_{\mathrm{T}}^{\mathrm{miss}}$ reconstruction starts with muons, electrons, followed by photons, then hadronically decaying τ-leptons, jets and finally the track soft term [37].

The missing transverse momentum can also be constructed using only inner detector tracks ($E_{\mathrm{T,trk}}^{\mathrm{miss}}$). This quantity provides a robust estimate of the missing transverse momentum which is less sensitive to the pile-up. However, since $E_{\mathrm{T,trk}}^{\mathrm{miss}}$ is based only on tracks left by charged particles it does not account for neutral particles or particles outside the ID.

The $E_{\mathrm{T}}^{\mathrm{miss}}$ reconstruction performance is derived by comparing $E_{\mathrm{T}}^{\mathrm{miss}}$ distributions in data and MC simulations for the same final state selection [38]. Systematics uncertainties in the $E_{\mathrm{T}}^{\mathrm{miss}}$ response and resolution are derived from these comparisons. $E_{\mathrm{T}}^{\mathrm{miss}}$ performance are obtained considering a final state without genuine $E_{\mathrm{T}}^{\mathrm{miss}}$ (as in $Z \rightarrow \mu\mu$ events) and a final state with genuine $E_{\mathrm{T}}^{\mathrm{miss}}$ (as in $W \rightarrow e\nu$ events). Figure 4.16 shows the $E_{\mathrm{T}}^{\mathrm{miss}}$ distributions in these two final states. For the $Z \rightarrow \mu\mu$ sample, data and MC simulations agree within 20% for the bulk of the $E_{\mathrm{T}}^{\mathrm{miss}}$ distribution, there are only large differences at high $E_{\mathrm{T}}^{\mathrm{miss}}$ values due to a mismodelling in $t\bar{t}$ events. In the $W \rightarrow e\nu$ sample with genuine $E_{\mathrm{T}}^{\mathrm{miss}}$, the level of agreement between data and MC simulation is 20% in the full missing transverse momentum range.

The $E_{\mathrm{T}}^{\mathrm{miss}}$ resolution is determined by the width of the distribution of the differences between the measured $E_{\mathrm{x(y)}}^{\mathrm{miss}}$ and the true missing transverse momentum vector $E_{\mathrm{T}}^{\mathrm{miss,\,true}}$:

$$\mathrm{RMS}_{\mathrm{x(y)}}^{\mathrm{miss}} = \mathrm{RMS}(E_{\mathrm{x(y)}}^{\mathrm{miss}} - E_{\mathrm{x(y)}}^{\mathrm{miss,\,true}}) \tag{4.12}$$

The $E_{\mathrm{T}}^{\mathrm{miss}}$ resolution is evaluated using dedicated samples from MC simulations ($W \rightarrow e\nu$, $W \rightarrow \mu\nu$ and $t\bar{t}$). Figure 4.17 shows the $E_{\mathrm{T}}^{\mathrm{miss}}$ resolution as a function of $E_{\mathrm{T}}^{\mathrm{miss,\,true}}$. The $E_{\mathrm{T}}^{\mathrm{miss}}$ resolution is very similar for $W \rightarrow e\nu$ and $W \rightarrow \mu\nu$ final states,

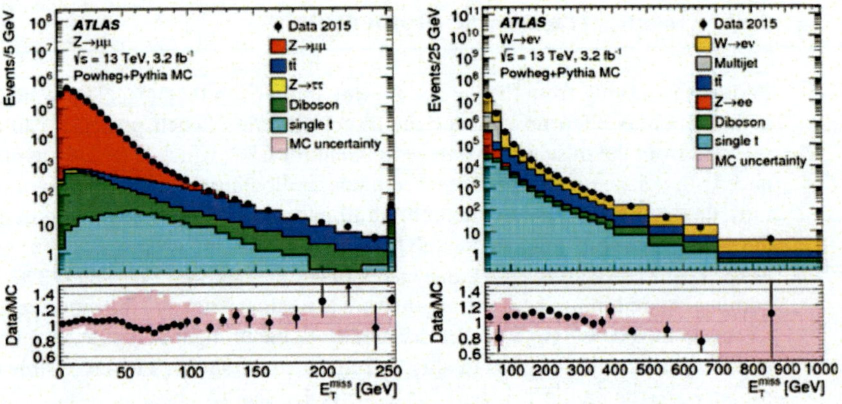

(a) (b)

Fig. 4.16 $E_{\mathrm{T}}^{\mathrm{miss}}$ distribution for $Z \to \mu\mu$ (**a**) and $W \to e\nu$ (**b**) events in data. The expectation from MC simulation is superimposed and includes all relevant background final states passing the event selection. The shaded areas indicate the total uncertainty for MC simulations, including the overall statistical uncertainty combined with systematic uncertainties from the p_{T} scale and resolution which are contributed by muons, electrons, jets and soft term. The last bin of each distribution includes the overflow. The respective ratios between data and MC simulations are shown in the bottom pad, with the shaded areas showing the total uncertainties for MC simulations [38]

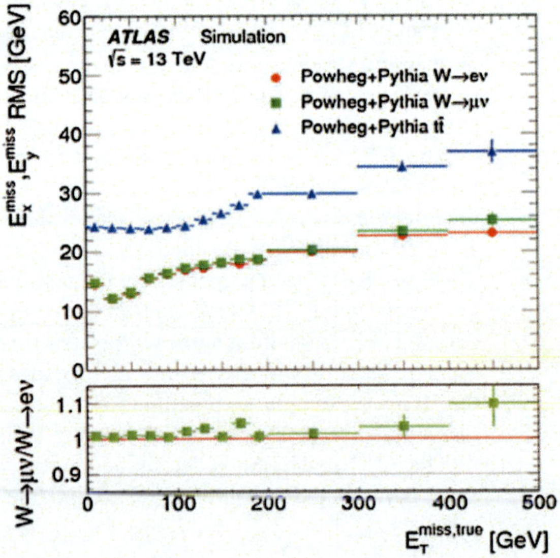

Fig. 4.17 $E_{\mathrm{T}}^{\mathrm{miss}}$ resolution measured by $\mathrm{RMS}_{\mathrm{x(y)}}^{\mathrm{miss}}$ as a function of the true missing transverse momentum $E_{\mathrm{T}}^{\mathrm{miss,\,true}}$ for the $W \to e\nu$, $W \to \mu\nu$ and $t\bar{t}$ samples from the MC simulations [38]

while it is poorer in $t\bar{t}$ final states. The poorer resolution in the last case is due to the presence of at least four jets with relatively low p_{T}. Additionally this topology is more sensitive to the fluctuations induced by the pile-up.

The uncertainties on the $E_{\mathrm{T}}^{\mathrm{miss}}$ are related only to the soft terms because the uncertainties on the hard terms are already included in the systematics related to the other physics objects. The $E_{\mathrm{T}}^{\mathrm{miss}}$ uncertainties come from the propagation of the energy scale and resolution uncertainties of the soft terms comparing data to MC simulation of events without genuine $E_{\mathrm{T}}^{\mathrm{miss}}$, such as $Z \rightarrow \mu\mu$ events. The difference between data and MC simulations is at most 10% at high $E_{\mathrm{T}}^{\mathrm{miss}}$.

4.5 Overlap Removal

During the object reconstruction, the information from the detector response could be wrongly reconstructed twice by multiple objects. To remove any double counting, a procedure known as *overlap removal* [39] is applied, with the aim of correctly identifying the true physical objects. Once all physical objects are fully reconstructed and calibrated, the algorithm is implemented. Individual overlaps are removed with a step-by-step procedure. Objects removed at a given step are not considered in the following. The step sequence is the following:

- **electron—muon**: if a combined muon shares an ID track associated to an electron, the electron is removed because it can be reconstructed if a muon radiates a photon.
- **electron—small-R jet**: if ΔR(small-R jet , e) < 0.2, the small-R jet is removed. This arises from the fact that the energy deposits in the EM calorimeter are always used as inputs for the jet reconstruction, regardless of the electron reconstruction result. For any surviving jets, if ΔR(small-R jet , e) < 0.4, the electron is removed because in this case the electron is considered incoming from semi-leptonic decays of heavy-flavour hadrons.
- **muon—small-R jet**: If ΔR(small-R jet, μ) < 0.2 or the muon ID track is ghost associated to the jet, then the jet is removed either if the jet has less than three associated tracks with $p_{\mathrm{T}} > 500$ MeV or both the following conditions are satisfied: the p_{T} ratio of the muon and jet is larger than 0.5 and the ratio of the muon p_{T} to the sum of p_{T} of tracks with $p_{\mathrm{T}} > 500$ MeV associated to the jet is larger than 0.7. This is because such a low track multiplicity can be compatible with radiation effects from the muon. For any surviving jets, if ΔR(small-R jet, μ) < 0.4, the muon is removed because the muon is considered as a product of a semi-leptonic heavy-flavour hadron decay.
- **electron—large-R jet**: if $\Delta R(e$, large-R jet) < 1.2, the large-R jet is removed.

In addition, an overlap removal procedure between VR track-jets is applied. Collinear VR track-jets are possible but the interplay between the VR track-jet reconstruction and the track-association in the b-tagging algorithm is questionable. To avoid this possible pathological case, a recommendation from the flavour tagging group is in

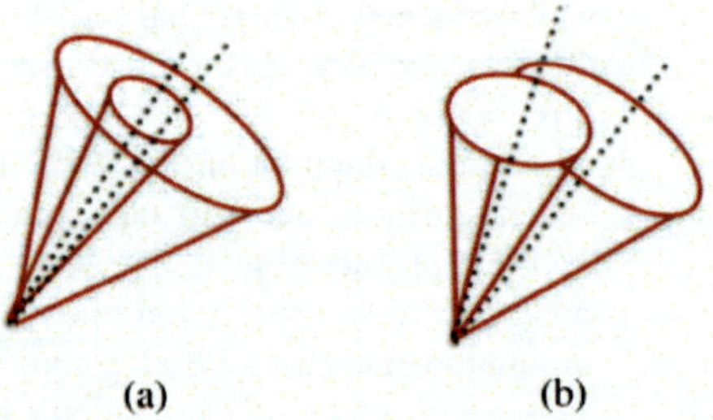

Fig. 4.18 **a** Example of events discarded when the overlap removal is applied. **b** Example of events considered when the OR is applied

place. The recommendation suggests to remove the events in which the following condition is satisfied:

$$\Delta R(\text{jet}_i, \text{jet}_j) < \min(R(\text{jet}_i), R(\text{jet}_j)) \quad (4.13)$$

where i runs on all the jets considered for the b-tagging algorithm and j runs on all the jets with a $p_{\text{T}} > 5$ GeV and with more than one track associated, with the caveat not to consider the case in which jet_i and jet_j coincide. With this recommendation all events in which both jet axes are reconstructed in both cones are excluded (Fig. 4.18a), while events with a topology similar to the one in Fig. 4.18b are considered.

The effects of this overlap removal has been studied also in the analysis presented in this thesis, focusing the attention on the 0-lepton channel. The percent range of discarded events is 2–6% and it depends on the energy of events (the higher the energy of the events, the higher probability that the event has two collinear VR track-jets).

References

1. Performance of the ATLAS inner detector track and vertex reconstruction in the high pile-up LHC environment, technical report. ATLAS-CONF-2012-042, CERN (2012). https://cds.cern.ch/record/1435196
2. Fruhwirth R (1987) Application of Kalman filtering to track and vertex fitting. Nucl Instrum Meth A 262:444. https://doi.org/10.1016/0168-9002(87)90887-4
3. ATLAS Collaboration (2017) Reconstruction of primary vertices at the ATLAS experiment in Run 1 proton-proton collisions at the LHC. Eur Phys J C77:332. https://doi.org/10.1140/epjc/s10052-017-4887-5. arXiv:1611.10235 [physics.ins-det]
4. ATLAS Collaboration (2015) Vertex reconstruction performance of the ATLAS detector at $\sqrt{s} = 13TeV$, technical report, ATL-PHYS-PUB-2015-026, CERN. https://cds.cern.ch/record/2037717
5. Collaboration ATLAS (2014) Electron reconstruction and identification efficiency measurements with the ATLAS detector using the 2011 LHC proton-proton collision data. Eur Phys J C 74:2941. https://doi.org/10.1140/epjc/s10052-014-2941-0 [hep-ex]
6. Collaboration ATLAS (2019) Electron reconstruction and identification in the ATLAS experiment using the 2015 and 2016 LHC proton-proton collision data at $\sqrt{s} = 13TeV$. Eur Phys J C 79:639. https://doi.org/10.1140/epjc/s10052-019-7140-6. arXiv:1902.04655 [hep-ex]
7. Lampl W et al. (2008) Calorimeter clustering algorithms: description and performance, ATLLARG- PUB-2008-002. https://cds.cern.ch/record/1099735

8. ATLAS Collaboration (2016) Electron efficiency measurements with the ATLAS detector using the 2015 LHC proton-proton collision data, ATLAS-CONF-2016-024. https://cds.cern.ch/record/2157687
9. ATLAS Collaboration (2016) Electron efficiency measurements in 2017 data and electron identification discriminating variables from 2016 data, technical report. https://atlas.web.cern.ch/Atlas/GROUPS/PHYSICS/PLOTS/EGAM-2018-002/index.html
10. ATLAS Collaboration (2016) Muon reconstruction performance of the ATLAS detector in proton-proton collision data at $\sqrt{s} = 13TeV$. Eur Phys J C 76:292. https://doi.org/10.1140/epjc/s10052-016-4120-y arXiv:1603.05598 [hep-ex]
11. ATLAS Collaboration, Muon ridentification and reconstruction efficiencies in full Run-2 dataset, technical report. https://atlas.web.cern.ch/Atlas/GROUPS/PHYSICS/PLOTS/MUON-2019-03/index.html
12. ATLAS Collaboration (2017) Topological cell clustering in the ATLAS calorimeters and its performance in LHC Run 1. Eur Phys J C 77:490. https://doi.org/10.1140/epjc/s10052-017-5004-5. arXiv:1603.02934 [hep-ex]
13. Cacciari M, Salam GP, Soyez G (2008) The $anti - k_t$ jet clustering algorithm. JHEP 04:063. https://doi.org/10.1088/1126-6708/2008/04/063. arXiv:0802.1189 [hep-ph]
14. Catani S, Dokshitzer YL, Seymour MH, Webber BR (1993) Longitudinally-invariant $k_{\perp}$- clustering algorithms for hadron-hadron collisions. Nucl Phys B 406:187. https://doi.org/10.1016/0550-3213(93)90166-M. https://cds.cern.ch/record/246812
15. Collaboration ATLAS (2017) Jet energy scale measurements and their systematic uncertainties in proton-proton collisions at $\sqrt{s} = 13TeV$ with the ATLAS detector. Phys Rev D 96:072002. https://doi.org/10.1103/PhysRevD.96.072002. arXiv:1703.09665 [hep-ex]
16. Cacciari M, Salam GP (2008) Pileup subtraction using jet areas. Phys Lett B 659:119. https://doi.org/10.1016/j.physletb.2007.09.077. arXiv:0707.1378 [hep-ph]
17. ATLAS Collaboration, Jet energy scale and uncertainties in 2015-2017 data and simulation, technical report. https://atlas.web.cern.ch/Atlas/GROUPS/PHYSICS/PLOTS/JETM-2018-006/
18. Collaboration ATLAS (2013) Jet energy resolution in proton-proton collisions at ps = 7 TeVrecorded in 2010 with the ATLAS detector. Eur Phys J C 73:2306. https://doi.org/10.1140/epjc/s10052-013-2306-0. arXiv:1210.6210 [hep-ex]
19. ATLAS Collaboration, Jet energy resolution in 2017 data and simulation, technical report. https://atlas.web.cern.ch/Atlas/GROUPS/PHYSICS/PLOTS/JETM-2018-005/
20. ATLAS Collaboration (2014) Tagging and suppression of pileup jets with the ATLAS detector,ATLAS-CONF-2014-018. https://cds.cern.ch/record/1700870
21. ATLAS Collaboration (2016) Jet mass reconstruction with the ATLAS Detector in early Run 2 data, ATLAS-CONF-2016-035. https://cds.cern.ch/record/2200211
22. ATLAS Collaboration, ATLAS combined mass resolution and W mass reconstruction performance, technical report. https://atlas.web.cern.ch/Atlas/GROUPS/PHYSICS/PLOTS/JETM-2017-002/
23. Collaboration ATLAS (2013) Performance of jet substructure techniques for large-R jets in proton-proton collisions at $\sqrt{s} = 7TeV$ using the ATLAS detector. JHEP 09:076. https://doi.org/10.1007/JHEP09(2013)076. arXiv:1306.4945 [hep-ex]
24. Krohn D, Thaler J, Wang L-T (2010) Jet Trimming. JHEP 02:084. https://doi.org/10.1007/JHEP02(2010)084. arXiv:0912.1342 [hep-ph]
25. ATLAS Collaboration, Jet substructire at very high luminosity, technical report. https://twiki.cern.ch/twiki/bin/view/AtlasPublic/JetSubstructureECFA2014
26. Collaboration ATLAS (2019) In situ calibration of large-radius jet energy and mass in 13 TeV proton-proton collisions with the ATLAS detector. Eur Phys J C 79:135. https://doi.org/10.1140/epjc/s10052-019-6632-8. arXiv:1807.09477 [hep-ex]
27. Krohn D, Thaler J, Wang L-T (2009) Jets with variable R. JHEP 06:059. https://doi.org/10.1088/1126-6708/2009/06/059. arXiv:0903.0392 [hep-ph]
28. ATLAS Collaboration (2017) Variable radius, exclusive-kT , and center-of-mass subjet reconstruction for Higgs ($\rightarrow b\bar{b}$) *tagging in ATLAS*, ATL-PHYS-PUB-2017-010. https://cds.cern.ch/record/2268678

29. ATLAS Collaboration (2014) Flavor tagging with track-jets in boosted topologies with the ATLAS detector, ATL-PHYS-PUB-2014-013. https://cds.cern.ch/record/1750681
30. ATLAS Collaboration (2016) Boosted Higgs ($\rightarrow b\bar{b}$) *Boson identification with the ATLAS detector at* $\sqrt{s} = 13TeV$, ATLAS-CONF-2016-039. https://cds.cern.ch/record/2206038
31. Cacciari M, Salam GP, Soyez G (2008) The catchment area of jets. JHEP 04:005. https://doi.org/10.1088/1126-6708/2008/04/005. arXiv:0802.1188 [hep-ph]
32. ATLAS Collaboration (2010) Performance of the ATLAS secondary vertex b-tagging algorithm in 7 TeV Collision Data, ATLAS-CONF-2010-042. https://cds.cern.ch/record/1277682
33. ATLAS Collaboration (2019) ATLAS b-jet identification performance and efficiency measurement with ($t\bar{t}$) *events in pp collisions at* $\sqrt{s} = 13TeV$. arXiv:1907.05120 [hep-ex]
34. ATLAS Collaboration (2017) Optimisation and performance studies of the ATLAS b-taggingalgorithms for the 2017-18 LHC run, ATL-PHYS-PUB-2017-013. https://cds.cern.ch/record/2273281
35. Collaboration ATLAS (2016) Performance of b-jet identification in the ATLAS experiment. JINST 11:P04008. https://doi.org/10.1088/1748-0221/11/04/P04008. arXiv:1512.01094 [hep-ex]
36. ATLAS Collaboration (2015) Expected performance of missing transverse momentum reconstructionfor the ATLAS detector at $\sqrt{s} = 13TeV$, ATL-PHYS-PUB-2015-023. https://cds.cern.ch/record/2037700
37. ATLAS Collaboration (2018) Emiss T performance in the ATLAS detector using 2015-2016 LHCpp collisions, ATLAS-CONF-2018-023. https://cds.cern.ch/record/2625233
38. Collaboration ATLAS (2018) Performance of missing transverse momentum reconstruction with the ATLAS detector using proton-proton collisions at $\sqrt{s} = 13TeV$. Eur Phys J C 78:903. https://doi.org/10.1140/epjc/s10052-018-6288-9. arXiv:1802.08168 [hep-ex]
39. Adams D et al. (2014) Recommendations of the physics objects and analysis Harmonisation study groups 2014, technical report. ATL-PHYS-INT-2014-018, CERN. https://cds.cern.ch/record/1743654

Chapter 5
Event Selection and Categorization

This chapter describes the selection of the $VH(b\bar{b})$ events in the boosted analysis. The first part discusses the criteria used to identify physics objects. The second part explains the selection of both the Higgs boson and vector boson candidates. In the final part all the details of the event categorization are illustrated.

5.1 Object Identification

The definition of the physics objects as well as the strategy of the *overlap removal* are provided in Sect. 4. In the following the list of all the objects, with their specific requirements, used in the $VH(b\bar{b})$ *boosted* analysis is presented. The description of the $E_{\mathrm{T}}^{\mathrm{miss}}$ object is skipped since the analysis does not apply any specific cut. In addition the b-tagging tool used to reconstruct the Higgs decay products is already shown in Sect. 4.3.7.

Leptons

The leptons used in the $VH(b\bar{b})$ analysis are electrons and muons.Two type of leptons are used in the analysis: *loose* leptons and *signal* leptons. The *loose* leptons are used to define the three main channels requiring exactly zero, one and two leptons. The *signal* leptons are a subset of the *loose* leptons with tighter identification and isolation requirements. In 1-lepton and 2-lepton channels at least one *signal* lepton is required to suppress the multi-jet background.

Loose electrons are required to have $p_{\mathrm{T}} > 7$ GeV and $|\eta| < 2.47$, to pass *Loose* identification requirement and *Fixed-Cut Loose* isolation (see Sect. 4.2). They have also to satisfy $|d_0/\sigma(d_0)| < 5$ and $|z_0 \sin(\theta)| < 0.5$ mm, where d_0 and z_0 are the transverse and longitudinal impact parameters defined relative to the primary vertex position and $\sigma(d_0)$ is the d_0 uncertainty. *Signal* electron in the 2-lepton channel are *loose* electron with a high p_{T} cut, $p_{\mathrm{T}} > 27$ GeV. Differently *signal* electrons in 1-lepton channel have tighter identification and isolation requirements with respect to

G. Di Gregorio, *A Beauty-ful Boson*, Springer Theses,
https://doi.org/10.1007/978-3-031-20013-7_5

Table 5.1 Electron selection requirements used in the $VH(b\bar{b})$ *boosted* analysis

Electron Selection	p_T	$\lvert\eta\rvert$	ID	$\lvert d_0/\sigma(d_0)\rvert$	$\lvert z_0 \sin(\theta)\rvert$	Isolation
Loose electron	>7 GeV	<2.47	*Loose*	<5	<0.5 mm	*Fixed-Cut Loose*
Signal electron in 2-lepton channel	>27 GeV	<2.47	*Loose*	<5	<0.5 mm	*Fixed-Cut Loose*
Signal electron in 1-lepton channel	>27 GeV	<2.47	*Tight*	<5	<0.5 mm	*Fixed-Cut High Pt Calo Only*

Table 5.2 Muon selection requirements used in the $VH(b\bar{b})$ *boosted* analysis.

Muon selection	p_T	$\lvert\eta\rvert$	ID	$\lvert d_0/\sigma(d_0)\rvert$	$\lvert z_0 \sin(\theta)\rvert$	Isolation
Loose muon	>7 GeV	<2.7	*Loose*	<3	<0.5 mm	*Fixed-Cut Loose*
Signal muon in 2-lepton channel	>27 GeV	<2.5	*Loose*	<3	<0.5 mm	*Fixed-Cut Loose*
Signal muon in 1-lepton channel	>25 GeV	<2.5	*Medium*	<3	<0.5 mm	*Fixed-Cut High Pt Track Only*

the *signal* electrons in 2-lepton channel in order to reject the multi-jet background. Table 5.1 summarizes the electron definitions used in the analysis.

Loose muons are required to have $p_T > 7$ GeV, $|\eta| < 2.7$, $|d_0/\sigma(d_0)| < 3$ and $|z_0 \sin(\theta)| < 0.5$. They have to pass *Loose* identification and *Fixed-Cut Loose* isolation requirements. *Signal* muons in the 2-lepton channel have the same requirements on the transverse and longitudinal impact parameters, together with the isolation and identification criteria of the *loose* muons. These muons must have $p_T > 27$ GeV and $|\eta| < 2.5$. *Signal* muons in 1-lepton channel have $p_T > 25$ GeV, $|\eta| < 2.5$, $|d_0/\sigma(d_0)| < 3$ and $|z_0 \sin(\theta)| < 0.5$. In addition they satisfy the *Medium* identification and the *Fixed-Cut High Pt Track Only* isolation criteria. Table 5.2 summarizes the muon definitions used in the analysis.

Jets

In the $VH(b\bar{b})$ analysis three different anti-k_t jet collections have been used: large-R jets (R=1.0), small-R jet ($R = 0.4$) and VR track-jets. The first two types of jets are reconstructed starting from the energy deposition in the calorimeter, while the track-jets are reconstructed from inner detector tracks.

The large-R jets are used to reconstruct the Higgs boson candidate in the high energy regime. Only large-R jets with $p_{\mathrm{T}} > 250$ GeV and $|\eta| < 2.0$ are considered. Due to the rule of thumb $\Delta R(jet_1, jet_2) \sim 2m/p_T$ where m and p_T are the mass and the transverse momentum of the large-R jet and ΔR is the angular separation between the decay products, the transverse momentum cut marks the point where the two b-quark jets are geometrically separated by $\Delta R = 1.0$.

To exploit the optimal large-R jet mass resolution over the full p_{T} range, the combined jet mass definition is used. A cut on the large-R jets mass m_J is applied, $m_J > 50$ GeV. The ATLAS Collaboration does not support the calibration in the low mass region due to the difference between data and simulation.

In the analysis two corrections are applied to the large-R jet to better set the scale and to improve the resolution of their energy and mass measurement. The large-R jets are first corrected to take into account the presence of muons from the b- or c-hadron decays. The muons are not included in the jet resulting in losses in the large-R jet energy. Therefore, a correction called *muon-in-jet*, which adds back the four-momentum of the muon associated to a large-R jet, is applied. When more than one muon is found, the one closest to the VR track-jet ghost-associated to the large-R jets is chosen.

The second correction called *Kinematic Fit* (KF) is applied only in the 2-lepton channel to improve the large-R jet energy resolution. The aim of this correction is to constrain the $ZH \rightarrow llb\bar{b}$ system to be balanced in the transverse plain. The KF includes also a constraint on the dilepton mass to the Z boson mass. This correction uses the energy resolution of the electrons and muons, which is typically 1%, to improve the large-R jet energy resolution, which is typically 10%.

To evaluate the improvement in resolution, the large-R jet mass m_J distributions before and after the corrections are fit with a Bukin function [1]. The resolution values σ correspond to the width of the fitted function. For the event selection of this analysis, the large-R jet mass resolution improves by 5% to 10% after the muon-in-jet correction, depending on the lepton channel. The KF brings an additional improvement in the 2-lepton channels of up to 40%. Figure 5.1 shows a comparison of the large-R jet mass when the additional corrections are applied to the jet energy scale in the 2-lepton channel.

In the $VH(b\bar{b})$ *boosted* analysis the small-R jets are used to build the $E_{\mathrm{T}}^{\mathrm{miss}}$, for the multi-jet estimate and for the event categorization. For the event categorisation they are required to have $p_{\mathrm{T}} > 30$ GeV and $|\eta| < 4.5$. To reduce the number of small-R jets originating from the pile-up interactions, the small-R jets are required to pass the JVT requirement if they are in the range $p_{\mathrm{T}} < 120$ GeV and $|\eta| < 2.5$.

In the analysis the VR track-jets are used as input of the b-tagging algorithm to correctly identify jets originating from the $H \rightarrow b\bar{b}$ decay. They are preferred over the standard calorimeter jets due to their higher efficiency to resolve objects in the high energy regions. Only central ($|\eta| < 2.5$) VR track-jets with $p_{\mathrm{T}} > 10$ GeV and with at least two tracks are considered.

Table 5.3 summarizes the jet collections and the respective kinematics cuts used in the analysis.

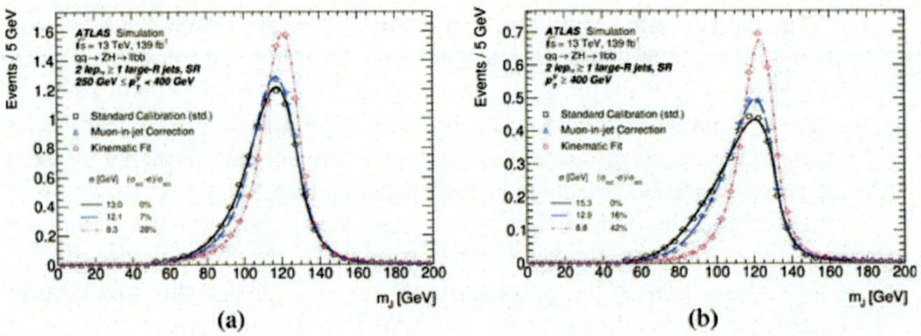

Fig. 5.1 Comparison of the large-R mass distributions when additional corrections are applied to the jet energy scale for the signal in the 2-lepton channel, 250 GeV $\leq p_T^V < 400$ GeV (a) and $p_T^V \geq 400$ GeV regions for the dominant $qq \rightarrow ZH$ contribution. The distributions are fitted with a Bukin function [1] and the resolution values, σ, correspond to the width of the fitted function [2]

Table 5.3 Jet collections and requirements

Jet collection	p_T	η
Large-R jets	>250 GeV	$\lvert\eta\rvert < 2.5$
Small-R jets	>30 GeV	$\lvert\eta\rvert < 4.5$
VR track-jets	>10 GeV	$\lvert\eta\rvert < 2.5$

5.2 Event Selections

Events are categorised into 0-, 1- and 2-lepton channels depending on the number of charged leptons (electrons or muons) to target the $ZH \rightarrow \nu\bar{\nu}b\bar{b}$, $WH \rightarrow l\nu b\bar{b}$ and $ZH \rightarrow llb\bar{b}$ signature, respectively.

The online selection of the 0-lepton channel relies on the E_T^{miss} trigger with increasing threshold, from 70 GeV to 110 GeV, for the increasing of luminosity during the Run 2. In the 1-lepton electron sub-channel events are selected by the a low-p_T threshold unprescaled single electron trigger. The same trigger as in the 0-lepton channel is used in the 1-lepton muon sub-channel since muons are not included in the online E_T^{miss} calculation. In the 2-lepton channel, the same trigger strategy as in the 1-lepton channel is adopted.

In the 0- and 2-lepton channels, the transverse momentum of the Z boson is reconstructed as E_T^{miss} and as the transverse momentum of the two leptons system, respectively. In the 1-lepton channel, the transverse momentum of the W boson is reconstructed as the vectorial sum of the E_T^{miss} and the lepton transverse momentum. In all the three lepton channels, the events are required to have the transverse momentum of the vector boson p_T^V grater than 250 GeV ($p_T^V \geq 250$ GeV). Events are split in two bins of p_T^V (250 GeV $\leq p_T^V < 400$ GeV, $p_T^V \geq 400$ GeV) to improve the analysis sensitivity. Events are also categorized in *signal regions* (SRs) and *control regions* (CRs). While the SR is the one expected to contain the larger fraction of signal, the

Table 5.4 Event selection for the three channels of the $VH(b\bar{b})$ *boosted* analysis

Selection	0 lepton channel	1 lepton channel		2 lepton channel	
		e sub-channel	μ sub-channel	*e* sub-channel	μ sub-channel
Trigger	$E_{\mathrm{T}}^{\mathrm{miss}}$	Single lepton	$E_{\mathrm{T}}^{\mathrm{miss}}$	Single lepton	$E_{\mathrm{T}}^{\mathrm{miss}}$
Leptons	0 *loose* lepton	1 *signal* lepton No second *loose* lepton		≥ 1 *signal* lepton 2 *loose* leptons	
$E_{\mathrm{T}}^{\mathrm{miss}}$	≥ 250 GeV	> 50GeV	–	–	
p_{T}^{V}	$p_{\mathrm{T}}^{V} \geq 250$ GeV				
Large-R jet	At least one large-*R* jet, $p_{\mathrm{T}} > 250$ GeV, $\|\eta\| < 2$				
Track-Jets	At least two track-jets, $p_{\mathrm{T}} > 10$ GeV, $\|\eta\| < 2.5$, matched to the leading large-*R* jet				
b-jets	Leading two track-jets matched to the leading large-*R* must be *b*-tagged				
m_J	> 50 GeV				
$\min[\Delta\phi(E_{\mathrm{T}}^{\mathrm{miss}},\text{jets})]$	$> 30°$	–			
$\Delta\phi(E_{\mathrm{T}}^{\mathrm{miss}},H_{\mathrm{cand}})$	$> 120°$	–			
$\Delta\phi\ (E_{\mathrm{T}}^{\mathrm{miss}}, E_{\mathrm{T,trk}}^{\mathrm{miss}})$	$< 90°$	–			
$\|\Delta y(V,H)\|$	–	$\|\Delta y(V,H)\| < 1.4$			
m_{ll}	–			66 GeV$< m_{ll} < 116$ GeV	
Lepton p_{T} imbalance	–			$(p_{\mathrm{T}}^{l_1} - p_{\mathrm{T}}^{l_2})/p_{\mathrm{T}}^{Z} < 0.8$	
Lepton flavor	–			Two lepton same flavour	
Lepton charge	–			Opposite sign muons	

CR is a background enriched region designed to evaluate the contribution and shape of one of the main backgrounds. In the 0- and 1-lepton channels, the SRs are defined by vetoing *b*-tagged track-jets outside the Higgs candidate jet. More details on the event categorization are reported in Sect. 5.3.

In the following paragraphs, a description of the Higgs candidate and channel-specific selections are provided. Table 5.4 summarizes the selection applied in each of the three channels.

5.2.1 Higgs Candidate Reconstruction and Selection

All the events in the three lepton channels are required to have at least one large-*R* jet. When more than one large-*R* jet is found in the event, the one with the highest transverse momentum is used as Higgs candidate jet. Events are also required to have at least two track-jets ghost matched to the Higgs candidate. All the track-jets in the events are required to pass the VR jet *overlap removal* procedure to avoid pathological cases to the *b*-tagging algorithm in which the two axes of the jets are reconstructed inside both cones (see Sect. 4.5). The *b*-tagging algorithm used in the analysis is the MV2 algorithm with the *b*-tagging efficiency of 70%.[1] The *b*-tagging algorithm is applied to the two VR track-jets with the highest transverse momentum

[1] The 70% efficiency of the *b*-tagging algorithm corresponds to a *c*-jet and light-flavour jet rejections equal to 9 and 304, respectively.

matched to the Higgs candidate and the events are categorized according to the number of b-tagged matched track-jets. Events with no b-tagged track-jets, or with exactly one b-tagged track-jet, or with exactly two b-tagged track-jets, compose the 0-tag, 1-tag and 2-tag categories, respectively. The analysis is performed in the 2-tag region because it is the region with the largest sensitivity in which the two b-tagged track-jets represents the two Higgs decay products. The 0-tag and 1-tag regions are used only in the 1-lepton channel to extract the QCD multi-jet contribution (see Sect. 6.3.7). The mass of the Higgs candidate is reconstructed using the invariant mass of the large-R jets m_J and the $m_J > 50$ GeV requirement is adopted. This request is applied before any large-R jet mass corrections (*muon-in-jet* correction and Kinematic Fit).

5.2.2 0-Lepton Channel Selection

In the 0-lepton channel a specific selection is defined to isolate events containing a Z boson decaying into a pair of neutrinos, in addition to the Higgs boson selection. Events passing the online selection are required to have no *loose* leptons and $E_{\mathrm{T}}^{\mathrm{miss}}$ ≥ 250 GeV.

In this channel the QCD multi-jet (MJ) events are a relevant background. The jet energy mis-measurements could generate fake $E_{\mathrm{T}}^{\mathrm{miss}}$ which tends to be aligned with the mis-measured jet. It is not possible to use the MC sample to estimate the MJ contributions because the simulated events do not populate the analysis phase space with enough statistics. The MC multi-jet samples are only used as cross-check once the MJ contribution is estimated with a data driven method. Multi-jet events are suppressed after applying angular cuts in the separation between small-R jets, $E_{\mathrm{T}}^{\mathrm{miss}}$, $E_{\mathrm{T,trk}}^{\mathrm{miss}}$ and Higgs candidate jet ($H_{\mathrm{cand.}}$):

- $\Delta\phi(E_{\mathrm{T}}^{\mathrm{miss}}, H_{\mathrm{cand.}}) > 120°$;
- $\Delta\phi(E_{\mathrm{T}}^{\mathrm{miss}}, E_{\mathrm{T,trk}}^{\mathrm{miss}}) < 90°$;
- $\min[\Delta\phi(E_{\mathrm{T}}^{\mathrm{miss}}, \text{small-}R \text{ jets})] > 30°$;

where $\Delta\phi(a, b)$ indicates the distance in the azimuthal angle between the two objects a and b. In the $\min[\Delta\phi(E_{\mathrm{T}}^{\mathrm{miss}}, \text{small-}R \text{ jets})]$ calculation, only small-R jets with p_{T} > 70 GeV geometrically outside[2] the Higgs candidate jet are considered. A detailed explanation on the $\min[\Delta\phi(E_{\mathrm{T}}^{\mathrm{miss}}, \text{small-}R \text{ jets})]$ cut and of the p_{T} threshold of the small-R jets is reported in Sect. 6.3.7. The values of the angular cuts are tuned in a way that the remaining fraction of MJ contamination is of the order of 1% of the signal.

The efficiency of each selection cut applied in 0-lepton channel has been studied using simulated signal samples. The efficiency, ϵ, is evaluated as the ratio of the number of events that pass a cut, N_{cut}, over the total number of generated events,

[2] Since the radius parameter of the Higgs candidate jet is $R = 1.0$, only small-R jets with an angular separation grater that 1.0 with the Higgs candidate are considered ($\Delta R(\text{small-}R \text{ jets}, H_{\mathrm{cand.}}) > 1.0$).

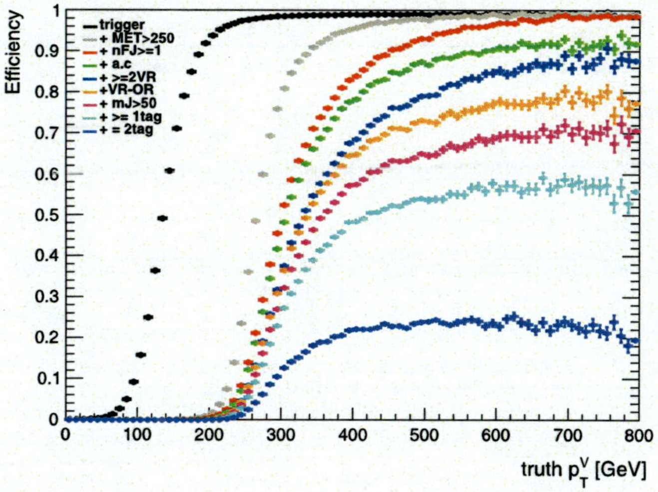

Fig. 5.2 Efficiency breakdown in 0-lepton channel. The cuts have been applied in cascade so in each step there is an additional request. Each curve with different colors shows the effect of cut cascade as described in the text

N_{all} ($\epsilon = N_{cut}/N_{all}$). Figure 5.2 shows the efficiency as a function of p_{T}^{V} evaluated at generator level (truth p_{T}^{V}). To study the efficiency the following cuts have been applied in cascade so in each step there is an additional request[3]:

- *trigger cut*: events that pass the trigger selection are required (black dots in Fig. 5.2);
- $E_{\mathrm{T}}^{\mathrm{miss}}$ *cut*: events with $E_{\mathrm{T}}^{\mathrm{miss}} \geq 250$ GeV are required (gray dots in Fig. 5.2);
- large-R *jet cut*: events with at least one large-R jet are required (red dots in Fig. 5.2);
- *anti-QCD cut*: events that pass the anti-QCD angular cuts are required (green dots in Fig. 5.2);
- *VR track-jets cut*: events with at least two VR track-jets ghost associated to the Higgs candidate jet are required (blue dots in Fig. 5.2);
- *VR overlap removal cut*: events that pass the VR overlap removal procedure are required (orange dots in Fig. 5.2);
- m_J *cut*: events with $m_J > 50$ GeV are required (magenta dots in Fig. 5.2);
- *1-tag cut*: events with at least one b-tagged track-jet are required (light blue dots in Fig. 5.2);
- *2-tag cut*: events with exactly two b-tagged track-jets are required (violet dots in Fig. 5.2).

[3] This means that the gray dots in Fig. 5.2 represent all the events that pass the trigger and $E_{\mathrm{T}}^{\mathrm{miss}}$ cuts divided by the total number of generated event.

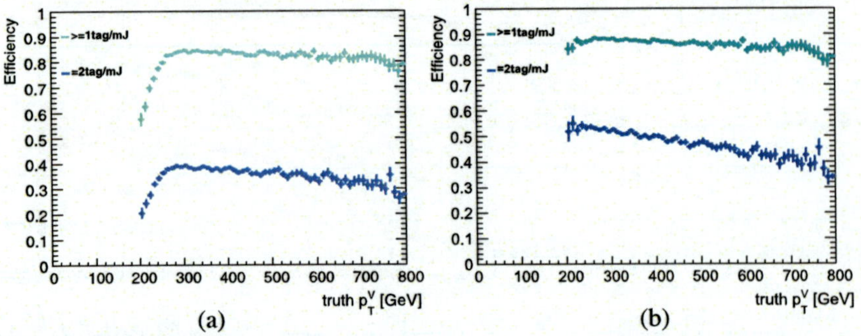

Fig. 5.3 **a** b-tagging efficiency in the 1-tag (light blue dots) and 2-tag (violet dots) region as a function of the truth transverse momentum of the vector boson. **b** b-tagging efficiency in the 1-tag (light blue dots) and 2-tag (violet dots) region as a function of the truth transverse momentum of the vector boson requiring also the VR track-jets reconstruct correctly the b-hadron

All the efficiencies have a turn on and then a flat trend. After applying all the cuts, only 20% of the signal events pass the 0-lepton event selection. As expected, the cut discarding most events is the b-tagging requirement. To understand better the impact of a single cut, it is needed to normalize the number of events that pass a specific cut to the number of events that pass the previous cut. In the following only the impact of the b-tagging algorithm is discussed. For completeness, all the other plots are reported in Appendix C. Figure 5.3a shows the number of events passing the event selection in the 1-tag (light blue dots) and 2-tag regions (violet dots) divided the number of events that pass all the event selection cuts except for the b-tagging request. The plot shows that both curves have a decreasing trend at high p_T^V values because the VR track-jets start to become closer and the b-tagging algorithm is less efficient. Knowing that the average b-tagging efficiency is 70%, in the 1-tag region the expected efficiency is 91%.[4] The observed efficiency in the 1-tag region is around 80% which is a lower than the expected value. Differently, the efficiency in the 2-tag region is 35%, while the expected efficiency is 49%. The discrepancy between the expected and the observed values has been further investigated. Figure 5.3b shows the efficiency in the 1-tag and 2-tag regions considering only events in which the VR track-jets reconstruct correctly the b-hadron. It is possible to know if the VR track-jet really contains the b-hadron looking at the generator level information of the simulated event. In this case the efficiency in the 1-tag region is approximately 91%, while in the 2-tag region is around 49%. Moreover, it has been also noticed that 25% of the signal events with at least three VR track-jets[5] have only one of the two leading VR track-jets which contains one b-hadron and the other VR track-jet

[4] In the 1-tag region either the leading or the sub-leading VR track-jets is b-tagged so the expected efficiency can be calculated subtracting to the unity the probability to have zero b-tagged track-jets which is 9% ($0.3 \times 0.3 = 0.09 = 9\%$).

[5] About 30% of the signal events have at least three VR track-jets.

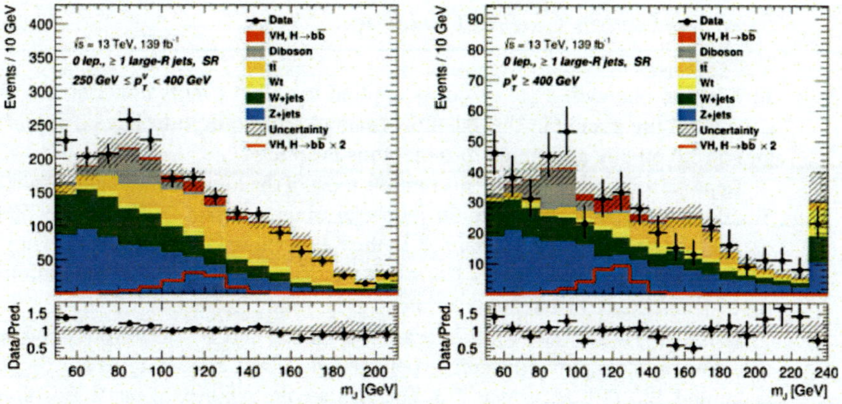

(a) (b)

Fig. 5.4 The invariant large-R jet mass m_J pre-fit distribution in the 0-lepton channel in the SR in the 250 GeV $\leq p_{\mathrm{T}}^{V} <$ 400 GeV (**a**) and $p_{\mathrm{T}}^{V} \geq$ 400 GeV (**b**) momentum ranges. The data are shown as black dots. The background contributions are shown as filled histograms. The Higgs boson signal is shown as a filled histogram on the top of the background contribution, and unstacked as an unfilled histogram, multiplied by a factor 2. The size of the combined statistical and systematic uncertainty for the sum of the signal and background is indicated by the hatched band. The highest bin in the distributions contains the overflow. The ratio of the data to the sum of the signal and background is shown in the lower panel

contains a light-hadron. The other b-hadron is contained in the third VR track-jet. This result shows that a possible way to recover the inefficiency of the b-tagging algorithm is to apply it considering the leading three VR track-jets instead of considering only the two leading VR track-jets.

The large-R jet mass m_J pre-fit distributions for data and simulated samples in the SR in the two p_{T}^{V} regions are shown in Fig. 5.4. The data are shown as black dots, while the signal and background contributions[6] are shown as filled histograms. The Higgs boson signal is also shown unstacked as an unfilled histogram multiplied by the factor indicated in the legend. The size of the combined statistical and systematic uncertainty for the sum of the signal and background is indicated by the hatched band. The highest bin in the distributions contains the overflow. The ratio of the data to the sum of the signal and background events is shown in the lower panel to highlight the agreement between data and MC simulation. The same style and convention are used also for the following plots in this thesis.

The dominant backgrounds of this channel are the Z+jets, W+jets and top processes. For this reason, a dedicated CR is set to model the top background (see Sect. 5.3.3).

[6] The contribution of the multi-jet background is not shown in the plots because, as discussed in the text, it is negligible after applying the event selection cuts. More information of the multi-jet background suppression can be found in Sect. 6.3.7.

5.2.3 1-Lepton Channel Selection

In 1-lepton channel, a set of cut is applied to select events containing a $W \rightarrow l\nu$ decay. All the events in the 1-lepton channel are required to have a *signal* lepton, and a veto on any additional *loose* leptons is applied.

Lepton isolation requirements remove most of the non-prompt lepton background. To additionally suppress the multi-jet background, a cut on $E_{\mathrm{T}}^{\mathrm{miss}}$ ($E_{\mathrm{T}}^{\mathrm{miss}} > 50$ GeV) is applied in the electron sub-channel. In the muon sub-channel, such cut is not applied because there are few events from the multi-jet background. More details on the multi-jet estimate in 1-lepton channel can be found in Sect. 6.3.7.

In order to reduce the contribution from the top and W+jets production, a further selection on the rapidity difference between the Higgs-candidate jet and the W boson is applied, $|\Delta y(W, H_{\mathrm{cand.}})| < 1.4$. To calculate the rapidity of the W boson it is necessary to fully reconstruct its momentum. Neglecting off-shell effects and W boson width, and assuming[7] $m^l = m^\nu = 0$ and $E_{\mathrm{T}}^{\mathrm{miss}} = p_T^\nu$, the longitudinal momentum of the neutrino is estimated by the following equation extracted constraining the lepton + neutrino system to have the W boson mass:

$$p_z^\nu = \frac{1}{2(p_{\mathrm{T}}^l)^2}\left[X p_z^l \pm E^l \sqrt{X^2 - \left[m_{\mathrm{T}W}^2 + 2 p_{\mathrm{T}}^l E_{\mathrm{T}}^{\mathrm{miss}} \cos\left(\Delta\phi(l, E_{\mathrm{T}}^{\mathrm{miss}})\right)\right]^2}\right] \tag{5.1}$$

with

$$X = m_W^2 + 2 p_{\mathrm{T}}^l E_{\mathrm{T}}^{\mathrm{miss}} \cos\left(\Delta\phi(l, E_{\mathrm{T}}^{\mathrm{miss}})\right) \tag{5.2}$$

where the l and ν superscripts represent the charged lepton and the neutrino, respectively, $\Delta\phi(l, E_{\mathrm{T}}^{\mathrm{miss}})$ is the azimuthal angle between the lepton and the $E_{\mathrm{T}}^{\mathrm{miss}}$, and $m_{\mathrm{T}W}$ is the transverse mass of the W boson defined as $m_{\mathrm{T}W} = \sqrt{2 p_{\mathrm{T}}^l E_{\mathrm{T}}^{\mathrm{miss}} (1 - \cos\left(\Delta\phi(l, E_{\mathrm{T}}^{\mathrm{miss}})\right))}$. This method leads to two solutions, the retained solution is the one that minimises the difference between the longitudinal boost of the W and Higgs bosons. The equation has imaginary solutions[8] if the discriminant is less zero which means that the transverse mass of the W boson $m_{\mathrm{T}W}$ is larger than the W boson mass m_W. In this particular case, the discriminant is set to zero which means setting the W transverse mass to the W mass ($m_{\mathrm{T}W} = m_W$). Preliminary studies show that the cut on the $|\Delta y(W, H_{\mathrm{cand.}})|$ discards 20% of the background events and 5% of signal events. The rapidity cut brings an improvement of 5% in the expected significance.

The efficiency of each selection cut used in the 1-lepton channel has been studied using MC signal samples. As in the 0-lepton channel, the efficiency is evaluated as the number of events that pass a cut over the total number of generated events. Figure 5.5 shows the efficiency of 1-lepton cuts as a function of truth p_{T}^V. To study

---

[7] The assumption of neglecting the lepton and neutrino masses is explained by the fact that the lepton and neutrino have masses much smaller that the W boson mass.

[8] The imaginary solutions are a consequence of the finite resolution of the detector.

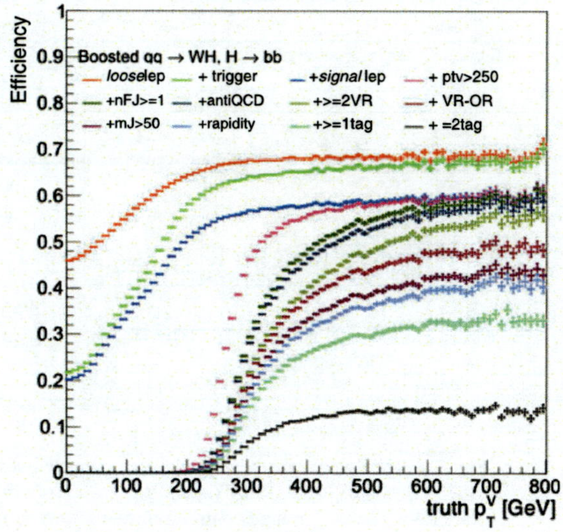

Fig. 5.5 Efficiency breakdown in 1-lepton channel. The cuts have been applied in cascade so in each step there is an additional request. Each curve with different colors shows the effect of cut cascade as described in the text

the efficiency the following cuts have been applied in cascade and in each step there is an additional request:

- *loose lepton cut*: events with a *loose* electron or *loose* muon are required (red dots);
- *trigger cut*: events that pass the trigger selection are required (lime dots);
- *signal lepton cut*: events with a *signal* electron or *signal* muon are required (blue dots);
- p_{T}^{V} *cut*: events with $p_{\mathrm{T}}^{V} > 250$ GeV are required (magenta dots);
- large-R *jet cut*: events with at least one large-R jet are required (dark green dots);
- *anti-QCD cut*: events that pass the anti-QCD cut ($E_{\mathrm{T}}^{\mathrm{miss}} > 50$ GeV in the electron sub-channel) are required (dark blue dots);
- *VR track-jets cut*: events with at least two VR track-jets ghost associated to the Higgs candidate jet are required (crocodile dots);
- *VR overlap removal cut*: events that pass the VR overlap removal procedure are required (purple dots);
- m_J *cut*: events with $m_J > 50$ GeV are required (violet dots);
- *rapidity cut*: events with $|\Delta y(W, H_{\mathrm{cand.}})| < 1.4$ are required (azure dots);
- *1-tag cut*: events with at least one b-tagged track-jet are required (turquoise dots);
- *2-tag cut*: events with exactly two b-tagged track-jets are required (black dots).

In the 1-lepton channel only 10% of the signal events has been selected and most of the events do not pass the *loose* lepton cut and the b-tagging requirements. The request of one electron or one muon in the final states reduces the number of the leptonic W events to 78% because the events with the W boson decays into $\tau + \nu$ would pass the event selection requirements only if the τ lepton decays leptonically. The observed efficiency is lower than this fraction because of the geometrical acceptance and lepton reconstruction efficiency.

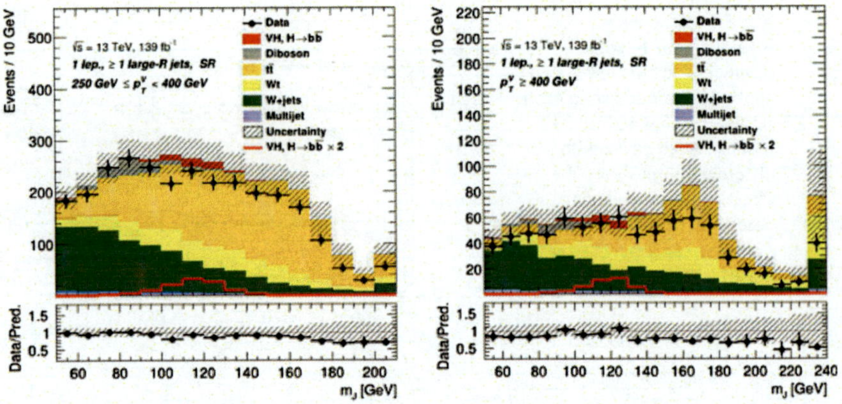

(a) (b)

Fig. 5.6 The invariant large-R jet mass m_J pre-fit distribution in the 1-lepton channel in the SR in the 250 GeV $\leq p_{\mathrm{T}}^{V} < 400$ GeV (**a**) and $p_{\mathrm{T}}^{V} \geq 400$ GeV (**b**) momentum ranges. The data are shown as black dots. The background contributions are shown as filled histograms. The Higgs boson signal is shown as a filled histogram on the top of the background contribution, and unstacked as an unfilled histogram, multiplied by a factor 2. The size of the combined statistical and systematic uncertainty for the sum of the signal and background is indicated by the hatched band. The highest bin in the distributions contains the overflow. The ratio of the data to the sum of the signal and background is shown in the lower panel

As in the 0-lepton channel the impact of each single cut has been studied. Almost all the events with truth $p_{\mathrm{T}}^{V} \geq 250$ GeV pass the *trigger cut* and less than 5% of the events are discarded. The *signal lepton cut* has an efficiency of 90% with a flat trend. With the *anti-QCD cut* less than 5% of signal events are discarded. The *rapidity cut* as an efficiency of 95%. The efficiencies of the large-R jet and VR track-jets related cuts have a similar trend in 0-lepton and 1-lepton channel. Also the efficiencies of the b-tagging cuts agree between 0-lepton and 1-lepton channel.

The large-R jet mass m_J pre-fit distributions for data and simulated samples in the SR in the two p_{T}^{V} regions are shown in Fig. 5.6. The dominant background contributions are the W+jets and top processes. Dedicated CRs are set to model the top backgrounds (see Sect. 5.3.3).

5.2.4 2-Lepton Channel Selection

In the 2-lepton channel, a Z boson decaying into two same flavour leptons (ee, $\mu\mu$) is reconstructed together with the Higgs candidate jet. Therefore selected events have exactly two same flavour *loose* leptons. Beside this request, at least one *signal* lepton is required. Moreover in the muon sub-channels, the two leptons must have opposite charge. The charge requirement is not applied to the di-electron events due to the higher rate of charge misidentification.

To reduce the Z+jets background, a cut on the rapidity difference between the Z and H bosons is applied. In this case, the four momentum of the Z boson is fully reconstructed using the information of the two leptons system. The same threshold as in 1-lepton channel is applied, $|\Delta y(Z, H_{\text{cand.}})| < 1.4$. To further suppress the non-resonant background, the invariant mass of the di-lepton system must be consistent with the Z boson mass, 66 GeV $< m_{ll} <$ 116 GeV.

The lepton-p_{T} imbalance $(p_{\text{T}}^{l_1} - p_{\text{T}}^{l_2})/p_{\text{T}}^{Z}$ is sensitive to the Z boson polarization which is found to be different between signal and Z+jets events [3]. The lepton-p_{T} imbalance can be used as a discriminant between signal and background events. The two leptons coming from the signal events usually have the same transverse momentum ($p_{\text{T}}^{l_1} \sim p_{\text{T}}^{l_2}$), while in case of $Z+$jets events the distribution of the lepton-p_{T} imbalance has a flat trend. To discard background events, the lepton-p_{T} imbalance is required to be less than 0.8.

The efficiency of each selection cut used in the 2-lepton channel has been studied using simulated signal samples. As before, the efficiency is evaluated as the number of events that pass a cut over the total number of generated events. Figure 5.7 shows the efficiency of the 2-lepton cuts as a function of the truth p_{T}^{V}. To study the efficiency, the following cuts have been applied in cascade and in each step there is an additional request:

- *loose lepton cut*: events with two *loose* electrons or two *loose* muons are required (red dots);
- *trigger cut*: events that pass the trigger selection are required (lime dots);
- m_{ll} *cut*: events with the invariant mass of the di-lepton system consistent with the Z boson mass, 66 GeV $< m_{ll} <$ 116 GeV. (blue dots);
- p_{T}^{V} *cut*: events with $p_{\text{T}}^{V} \geq 250$ GeV are required (magenta dots);
- large-R *jet cut*: events with at least one large-R jet are required (light blue dots);
- *VR track-jets cut*: events with at least two VR track-jets ghost associated to the Higgs candidate jet are required (green dots);
- *VR overlap removal cut*: events that pass the VR overlap removal procedure are required (gray dots);
- m_J *cut*: events with $m_J > 50$ GeV are required (violet dots);
- *rapidity cut*: events with $|\Delta y(Z, H_{\text{cand.}})| < 1.4$ are required (golden dots);
- *lepton-p_{T} imbalance cut*: events with $(p_{\text{T}}^{l_1} - p_{\text{T}}^{l_2})/p_{\text{T}}^{Z} < 0.8$ are required (light brown dots);
- *1-tag cut*: events with at least one b-tagged track-jet are required (brown dots);
- *2-tag cut*: events with exactly two b-tagged track-jets are required (light red dots).

As in the 1-lepton channel only 10% of the simulated signal events pass the full cut cascade. The request of two *loose* leptons in the final state removes 40–50% of the events. The shape of the efficiency of the *trigger cut* has a discontinuity because in the muon sub-channel the $E_{\text{T}}^{\text{miss}}$ trigger is used from $p_{\text{T}}^{V} > 150$ GeV. Events with a p_{T}^{V} below 150 GeV have been selected using a single muon trigger. All the di-muons events used in the analysis have a $p_{\text{T}}^{V} \geq 250$ GeV so the single muon trigger is never

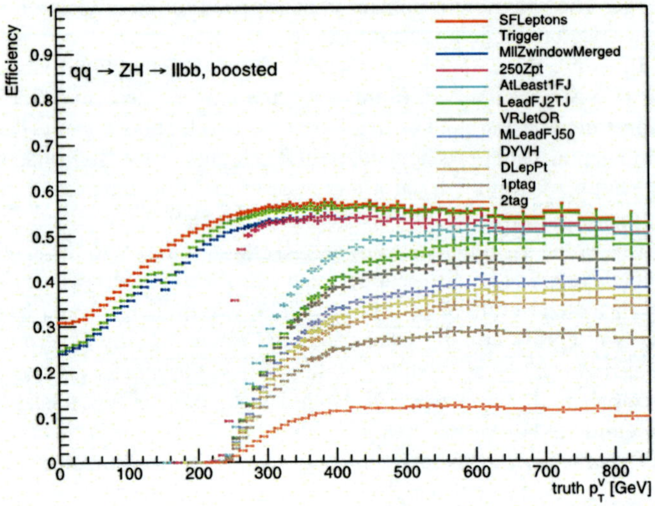

Fig. 5.7 Efficiency breakdown in 2-lepton channel. The cuts have been applied in cascade so in each step there is an additional request. Each curve with different colors shows the effect of cut cascade as described in the text

applied. After applying the *loose lepton* and *trigger cuts*, it has been tested the request of having two opposite muons in the muon-sub channel. This cut is fully efficient and for this reason is not reported in Fig. 5.7. The m_{ll} *cut* is almost fully efficient, only 2% of the events are discarded. After a sharp turn-on, the p_T^V *cut* is fully efficient. With the *lepton-p_T imbalance cut* less than 10% of the events are removed. The efficiency of the jets related cuts and *b*-tagging requirements have been compared among the three lepton channels and the results of the 2-lepton channel are in agreement with the one of the 0-lepton and 1-lepton channel. Also the rapidity cut has the same trend in 1- and 2-lepton channels.

The large-*R* jet mass pre-fit distributions for data and simulated sample in the SR in the two p_T^V regions are shown in Fig. 5.8. In this channel, the dominant background is the Z+jets process.

5.3 Event Categorization

After applying the requirements described above, the events in the three lepton channels are categorized depending on the p_T^V. In 0-lepton and 1-lepton channels, the events are further categorized depending on the number of small-*R* jets and on the number of *b*-tagged track-jets outside the Higgs candidate jets. At the end of this section there is a summary of the analysis region definition.

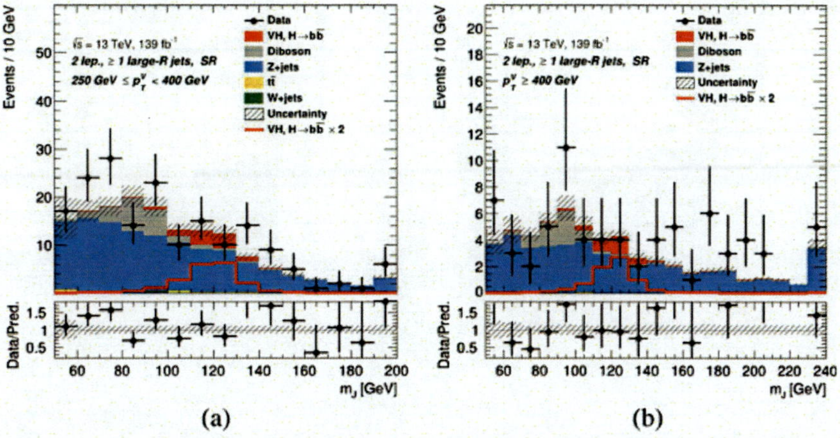

Fig. 5.8 The invariant large-R jet mass m_J pre-fit distribution in the 2-lepton channel in the SR in the 250 GeV $\leq p_T^V <$ 400 GeV (**a**) and $p_T^V \geq$ 400 GeV (**b**) momentum ranges. The data are shown as black dots. The background contributions are shown as filled histograms. The Higgs boson signal is shown as a filled histogram on the top of the background contribution, and unstacked as an unfilled histogram, multiplied by a factor 2. The size of the combined statistical and systematic uncertainty for the sum of the signal and background is indicated by the hatched band. The highest bin in the distributions contains the overflow. The ratio of the data to the sum of the signal and background is shown in the lower panel

5.3.1 p_T^V Splitting

The events are categorized depending on the p_T^V because the phase-space with high signal-to-background ratio is at high values of the vector boson transverse momentum. Moreover, BSM effects may be more pronounced in the high-p_T region. Two regions are considered: a medium energy region and a high energy region, 250 GeV $\leq p_T^V <$ 400 GeV and $p_T^V \geq$ 400 GeV, respectively. These p_T^V intervals are coherent with the cuts used in the STXS categorization (see Sect. 1.5). Figure 5.9 shows the p_T^V pre-fit distributions in the SR in the three lepton channels.

5.3.2 Signal Region Splitting

In 0- and 1-lepton channels the SR is defined by requiring to have zero b-tagged track-jets outside the Higgs candidate jet in order to enhance the top background rejection. It is possible to further discriminate between the top process and the signal using the jet multiplicity of the two processes. The difference in jet multiplicity can be easily deduced from the leading order Feynman diagrams for the $t\bar{t}$ and WH processes shown in Fig. 5.10. The $t\bar{t}$ events passing the event selection are mainly

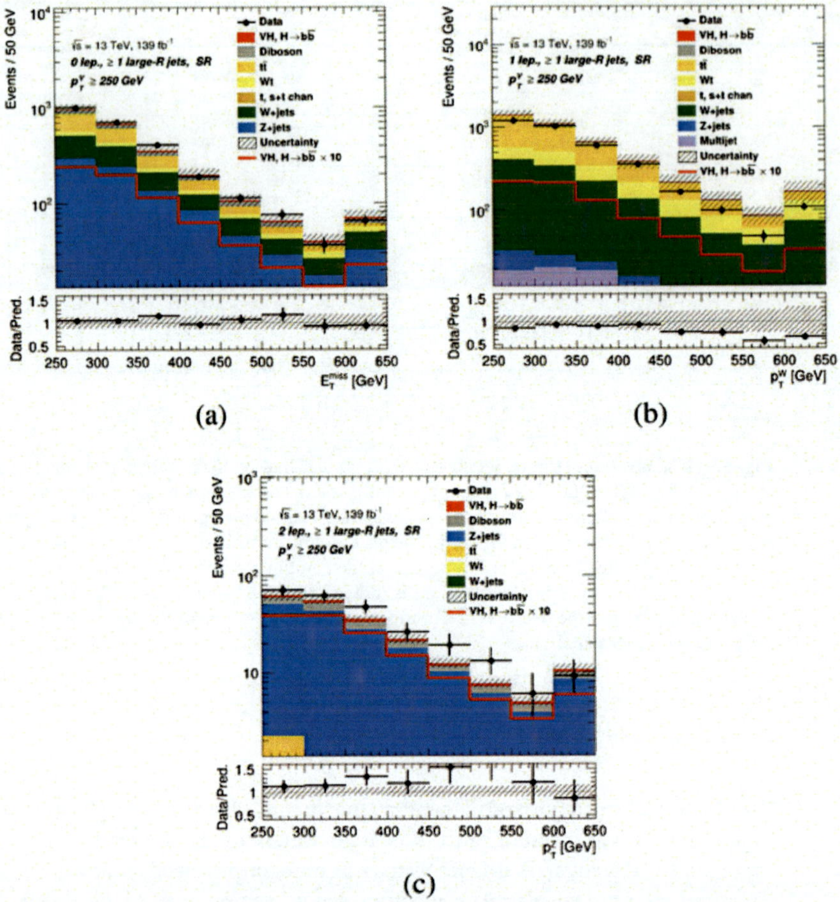

Fig. 5.9 The p_T^V pre-fit distributions in 0-lepton (**a**), 1-lepton (**b**) and 2-lepton (**c**) SRs for $p_T^V \geq 250$ GeV. The data are shown as black dots. The background contributions are shown as filled histograms. The Higgs boson signal is shown as a filled histograms on top of the background contribution, and unstacked as an unfilled histogram, multiplied by a factor 10. The size of the combined statistical and systematic uncertainty for the sum of the signal and background is indicated by the hatched band. The highest bin in the distributions contains the overflow. The ratio of the data to the sum of the signal and background is shown in the lower panel

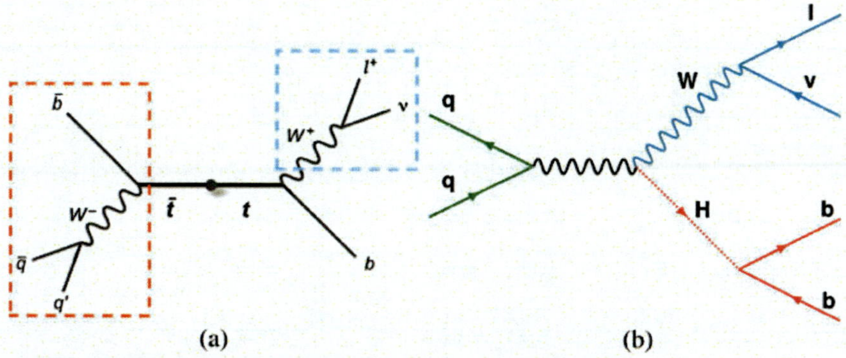

Fig. 5.10 Leading order Feynmam diagram for the $t\bar{t}$ (**a**) and WH (**b**) processes. The red box in the Feynman diagram of the $t\bar{t}$ process indicates the decay products of a hadronically decaying top quark which can be wrongly selected as Higgs candidate jet. The light blue box includes the decay products of a W boson which is correctly reconstructed as V boson candidate

constituted by events in which one W boson decays leptonically and the other one decays hadronically. The W boson that decays leptonically is expected to be identified by the vector boson selection (light blue box in Fig. 5.10a). One of the two jets from the hadron decay of the W boson could be mis-identified as a b-jet. In the red box, this jet together with the b-quark of the top decay can be wrongly selected as Higgs candidate (red box in Fig. 5.10a). As a result, the event has an additional b-jet outside the Higgs candidate jet.

An additional way to discriminate top events from signal events is study the hadron activity of the event because $t\bar{t}$ events are characterised by more activity. The angular distance between the Higgs candidate jet and the small-R jets is studied to avoid using jets associated to the Higgs candidate jet.[9] Figure 5.11 shows the angular distance between the Higgs candidate and the leading small-R jet for signal VH events in 1-lepton signal region. The request $\Delta R > 1$ ensures that the small-R jet is not matched to the Higgs candidate jet.

To assess the jet activity, the p_{T} distribution of the leading small-R jet not matched to the Higgs candidate jet is studied. It is expected that the jets produced by $t\bar{t}$ events have higher p_{T} with respect to the jets produced by signal events. Figure 5.12 shows the p_{T} distribution of the leading small-R jet not matched to the Higgs candidate jet in 1-lepton SR, 250 GeV $\leq p_{\mathrm{T}}^{V} < 400$ GeV, for the major processes. The distribution of each process is normalized to the total number of events of the process that pass the event selection. Looking at the plot, the distribution of the $t\bar{t}$ events has a maximum at $p_{\mathrm{T}} \sim 30$ GeV, while the maximum of the distribution of signal events is around $p_{\mathrm{T}} \sim 20$ GeV. To discriminate signal from $t\bar{t}$ events, it is required that the small-R jets

[9] Small-R jets are used in these studies instead of VR track-jets in order to have an easier estimate of the systematics uncertainties.

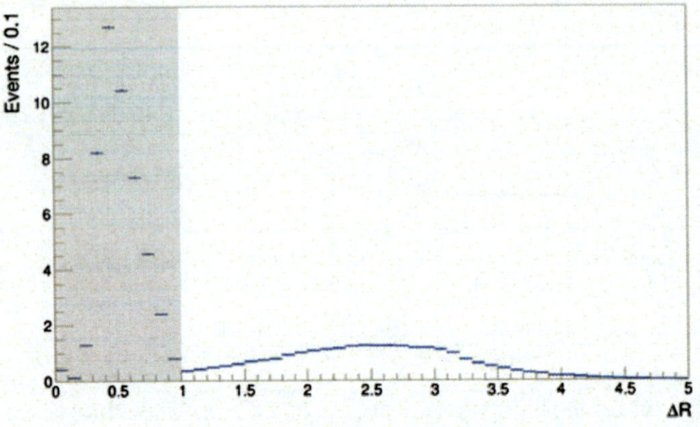

Fig. 5.11 Distribution of the angular distance between the Higgs candidate and the leading small-R jet for VH signal events in 1-lepton SR, 250 GeV $\leq p_{\mathrm{T}}^{V} < 400$ GeV. The gray area represents all the signal events in which the selected small-R jet is matched to the Higgs candidate

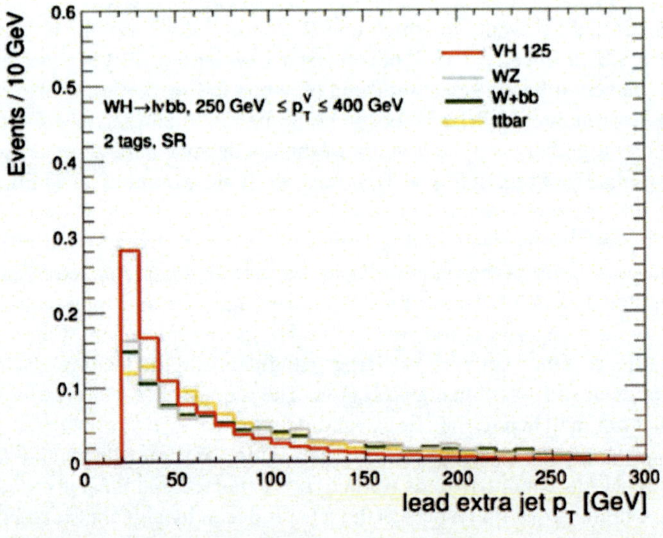

Fig. 5.12 p_{T}^{V} distribution of the leading small-R jet outside the Higgs candidate jet in 1-lepton SR, 250 GeV $\leq p_{\mathrm{T}}^{V} < 400$ GeV, for the main processes. The distribution of each process is normalized to the total number of events of the process that pass the event selection

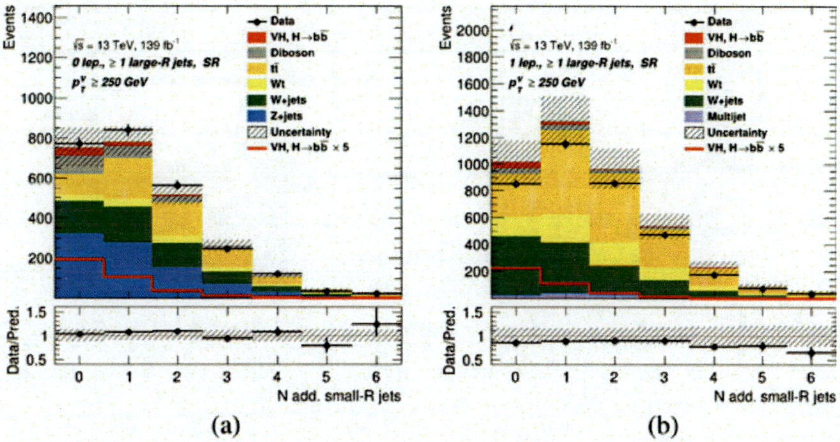

Fig. 5.13 Number of small-R jets non-matched to the Higgs candidate jet distribution in 0-lepton (**a**) and 1-lepton (**b**) SRs for $p_T^V \geq 250$ GeV. The data are shown as black dots. The background contributions are shown as filled histograms. The Higgs boson signal is shown as a filled histograms on top of the background contribution, and unstacked as an unfilled histogram, multiplied by a factor 5. The size of the combined statistical and systematic uncertainty for the sum of the signal and background is indicated by the hatched band. The highest bin in the distributions contains the overflow. The ratio of the data to the sum of the signal and background is shown in the lower panel

outside the Higgs candidate jet must have $p_T > 30$ GeV. This p_T cut is also aligned with one applied on the small-R jets in the STXS framework (see Sect. 1.5).

The final step is to apply a cut on the jet multiplicity. Figure 5.13 shows the distribution of the number of small-R jets outside the Higgs candidate in the 0-lepton and 1-lepton channel. In both channels the distribution of $t\bar{t}$ events has different shape with respect to the shape of the signal events which has a peak at zero.

Requiring zero small-R jets outside the Higgs candidate jet it is possible to discard about 60(70%) of backgrounds events and 35% of signal events in 0-lepton (1-lepton) channel. In the analysis the region with zero small-R jets outside the Higgs candidate jet is called *high purity signal region* (HP SR) because it is the region with highest signal-to-background ratio and less top events. To avoid signal loss, events in the SR with one or more small-R jets are used to define the *low purity signal region* (LP SR). This categorization of the SR is only applied in the 0- and 1-lepton channels. It is not applied in the 2-lepton channel because the $t\bar{t}$ process is an almost negligible background. Preliminary studies show that the splitting of the SR brings a 30% (17%) gain in the expected significance in 1-lepton (0-lepton) channel. The improvement in the 1-lepton channel is bigger with respect to the one in 0-lepton channel because in 1-lepton channel the fraction of $t\bar{t}$ events is higher than in 0-lepton channel.

The large-R jet mass m_J pre-fit distributions for data and simulated sample in the HP SR and LP SR in the two p_T^V regions in 0-and 1-lepton channels are shown in Figs. 5.14 and 5.15.

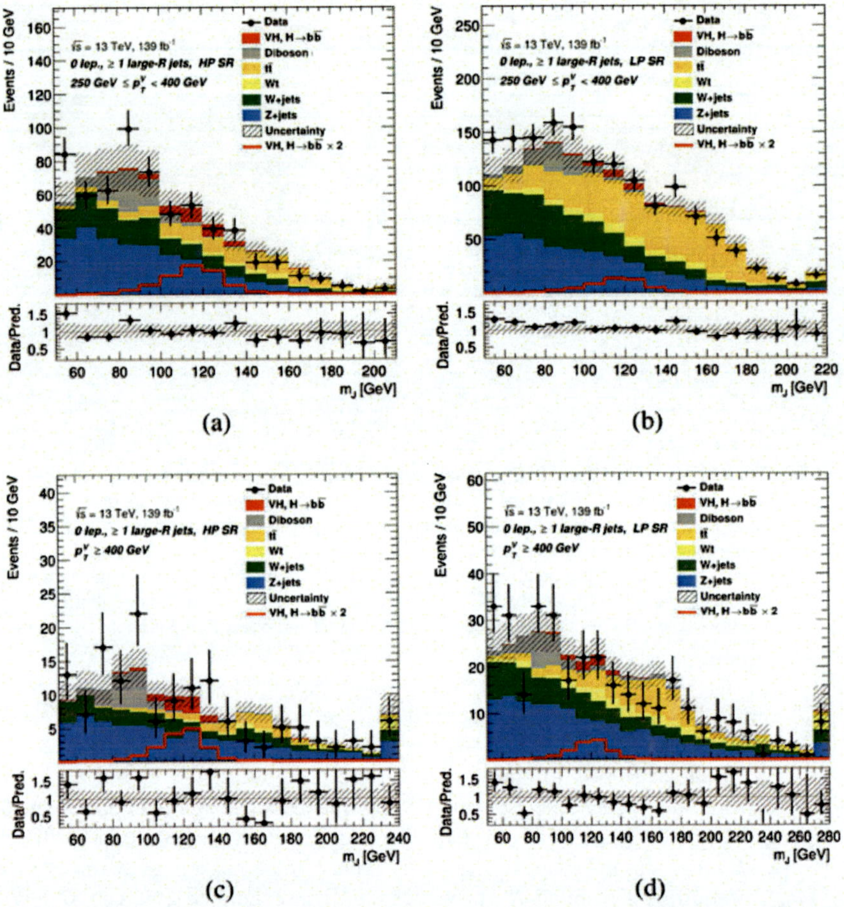

Fig. 5.14 The invariant large-R jet mass m_J pre-fit distributions in the 0-lepton channel in the HP SR (left) and LP SR (right) in the 250 GeV $\leq p_{\mathrm{T}}^{V} < 400$ GeV (top) and $p_{\mathrm{T}}^{V} \geq 400$ GeV (bottom) region. The data are shown as black dots. The background contributions are shown as filled histograms. The Higgs boson signal is shown as a filled histogram on the top of the background contribution, and unstacked as an unfilled histogram, multiplied by a factor 2. The size of the combined statistical and systematic uncertainty for the sum of the signal and background is indicated by the hatched band. The highest bin in the distributions contains the overflow. The ratio of the data to the sum of the signal and background is shown in the lower panel

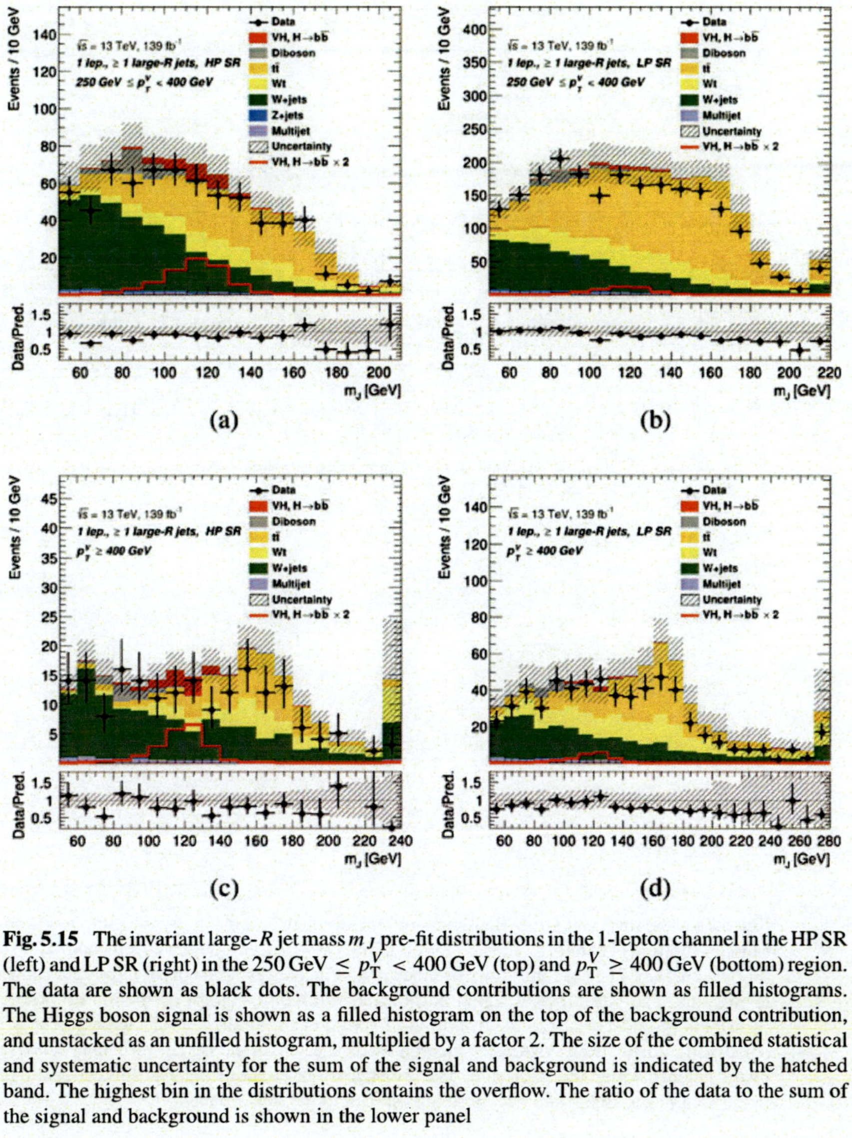

Fig. 5.15 The invariant large-R jet mass m_J pre-fit distributions in the 1-lepton channel in the HP SR (left) and LP SR (right) in the 250 GeV $\leq p_T^V < 400$ GeV (top) and $p_T^V \geq 400$ GeV (bottom) region. The data are shown as black dots. The background contributions are shown as filled histograms. The Higgs boson signal is shown as a filled histogram on the top of the background contribution, and unstacked as an unfilled histogram, multiplied by a factor 2. The size of the combined statistical and systematic uncertainty for the sum of the signal and background is indicated by the hatched band. The highest bin in the distributions contains the overflow. The ratio of the data to the sum of the signal and background is shown in the lower panel

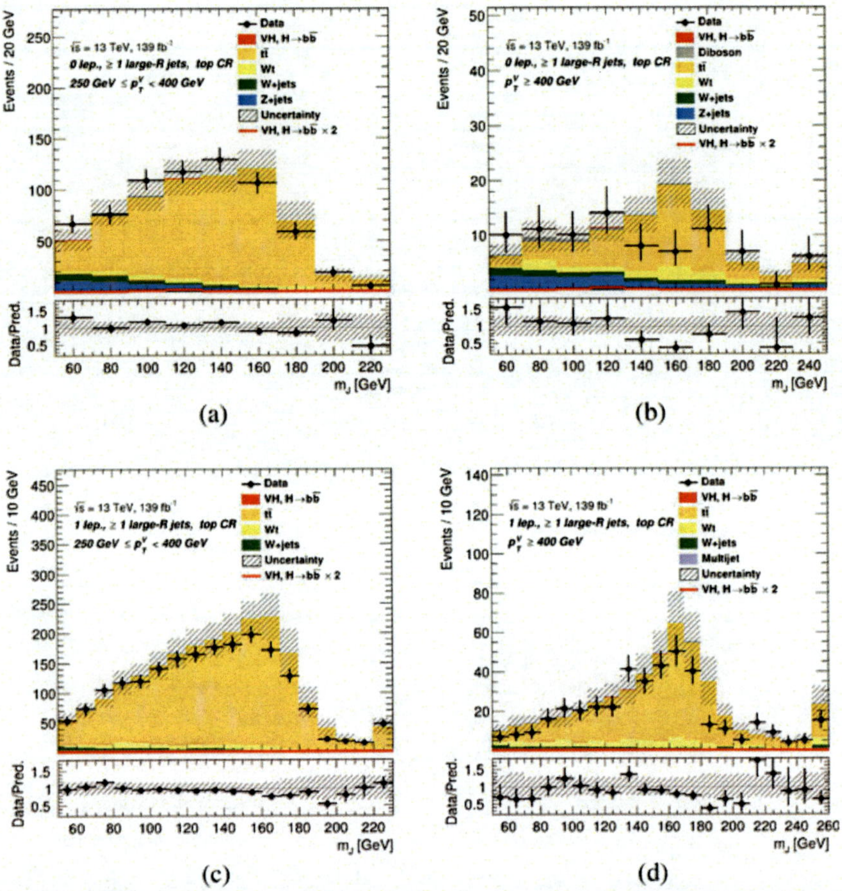

Fig. 5.16 The invariant large-R jet mass m_J pre-fit distributions in the 0-lepton (top) and 1-lepton channel (bottom) channel in the CR in the 250 GeV $\leq p_T^V < 400$ GeV (left) and $p_T^V \geq 400$ GeV (bottom) region. The data are shown as black dots. The background contributions are shown as filled histograms. The Higgs boson signal is shown as a filled histogram on the top of the background contribution, and unstacked as an unfilled histogram, multiplied by a factor 2. The size of the combined statistical and systematic uncertainty for the sum of the signal and background is indicated by the hatched band. The highest bin in the distributions contains the overflow. The ratio of the data to the sum of the signal and background is shown in the lower panel

5.3.3 Control Region Definition

The $t\bar{t}$ process is one of the main background in the 0- and 1-lepton channel and a way to further study it is building a control region. In the 0-lepton and 1-lepton channel, the CRs are defined requiring at least one b-tagged track-jet outside the Higgs candidate jet.[10] As illustrated in Fig. 5.10a, the $t\bar{t}$ decay can give at least one stand-alone b-quark outside the Higgs candidate jet.

Figure 5.16 shows the m_J pre-fit distributions in the CRs in 0- and 1-lepton channels. The plots confirm that most of the events in the CRs are from the $t\bar{t}$ background process. Due to the low statistic in the 0-lepton channel a coarser binning is chosen with respect to the 1-lepton channel. Moreover, comparing the data to the MC prediction, the distributions show a mis-modelling at around the top mass.

In the 0-lepton channels 80% (56%) of the events in the CRs are from the top process in the 250 GeV $\leq p_{\mathrm{T}}^{V} < 400$ GeV ($p_{\mathrm{T}}^{V} \geq 400$ GeV) region, while in 1-lepton channel the fraction of $t\bar{t}$ events is ~90%. In 1-lepton channel the fraction of $t\bar{t}$ events in the CRs is higher than in the 0-lepton channel because the signal final state is more similar to the $t\bar{t}$ final state.

5.3.4 Summary of the Signal and Control Regions

According to the event categorization described in the previous sub-sections, in the analysis ten SRs and four CRs are considered. The SRs and CRs are summarized in Table 5.5.

Table 5.5 Summary of the definition of the analysis regions. Regions with relatively large signal purity are marked with the label SR. Background enriched regions are marked with the label CR

Channel	Categories					
	$250 < p_T^V < 400$ GeV			$p_T^V > 400$ GeV		
	0 add. b track-jets		≥ 1 add. b track-jets	0 add. b track-jets		≥ 1 add. b track-jets
	0 add. small-R jets	$\geq$ 1add. small-R jets		0 add. small-R jets	$\geq$ 1add. small-R jets	
0-lepton	SR	SR	CR	SR	SR	CR
1-lepton	SR	SR	CR	SR	SR	CR
2-lepton	SR			SR		

[10] In this case, the b-tagged track-jet is considered outside the Higgs candidate jet if it is not ghost-matched to the Higgs candidate jet.

References

1. Bukin AD (2007) Fitting function for asymmetric peaks. arXiv:0711.4449 [physics.data-an]
2. Aad G et al. (2020) Measurement of the associated production of a Higgs boson decaying into b-quarks with a vector boson at high transverse momentum in pp collisions at $\sqrt{s} = 13TeV$ with the ATLAS detector. arXiv:2008.02508 [hep-ex]
3. Goncalves D, Nakamura J (2018) Role of the Z polarization in the $H \to b\bar{b}$ measurement. Phys Rev D 98:093005. https://doi.org/10.1103/PhysRevD.98.093005. arXiv:1805.06385 [hep-ph]

Chapter 6
Systematic Uncertainties

This chapter is dedicated to the sources of systematics uncertainties that affect the $VH(b\bar{b})$ measurement. A particular emphasis is given to the evaluation of the modelling uncertainties since they are analysis specific.

6.1 Introduction

The uncertainties that affect the measurements can be divided into statistical uncertainties and systematics uncertainties. The full statistical model used in the measurement will be described in Chap. 7. While statistical uncertainties are the result of stochastic fluctuations in the data sample, systematics uncertainties are associated with the calibration and resolution of the measurement apparatus, and the assumptions made to model signal and background events in the measurement. The source of systematics can be divided in two groups: uncertainties related to the experimental setup and uncertainties related to the modelling of simulated data. In the following sections these categories will be described in more details.

6.2 Experimental Uncertainties

Experimental uncertainties are all the systematic uncertainties related to the ability to simulate detector effects and experimental resolutions. They follow the official ATLAS recommendations and they are analysis independent. The dominant experimental uncertainties originate from the large-R jets and from b-tagging simulation-to-data efficiency correction factors. All the experimental uncertainties considered in the analysis are briefly described in the following and summarised in Table 6.1. The first column of Table 6.1 shows the name of the systematic uncertainty used in the fit. The name will be used in Chap. 8 to understand fit results.

G. Di Gregorio, *A Beauty-ful Boson*, Springer Theses,
https://doi.org/10.1007/978-3-031-20013-7_6

Table 6.1 Summary of the experimental systematic uncertainties applied in the $VH(b\bar{b})$ analysis. The first column shows the name of the systematic uncertainty used in the fit

Systematic uncertainty name	Description
Event	
LUMI_2015_2018	Uncertainty on total integrated luminosity
PRW_DATASF	Uncertainty on pile-up modelling
Electrons	
EL_EFF_Trigger_TOTAL	Trigger efficiency uncertainty
EL_EFF_Reco_TOTAL	Reconstruction (reco.) efficiency uncertainty
EL_EFF_ID_TOTAL	Identification (id.) efficiency uncertainty
EL_EFF_Iso_TOTAL	Isolation efficiency uncertainty
EG_SCALE_ALL	Energy scale uncertainty
EG_RESOLUTION_ALL	Energy resolution uncertainty
Muons	
MUON_EFF_RECO_STAT(SYS)	Reconstruction and ID efficiency uncertainty for muons with $p_T > 15$ GeV
MUON_EFF_RECO_STAT(SYS)_LOWPT	Reco. and ID efficiency uncertainty for muons with $p_T < 15$ GeV
MUON_EFF_ISO_STAT(SYS)	Isolation efficiency uncertainty
MUON_TTVA_STAT(SYS)	Track-to-vertex association efficiency uncertainty
MUON_SCALE	Momentum scale uncertainty
MUON_SAGITTA_RHO(RESBIAS)	Momentum scale uncertainty to cover charge-dependent local misalignment effects
MUON_ID(MS)	Momentum resolution uncertainty of the inner detector (muon spectrometer)
Small-R jets	
JET_BJES_Response	Energy scale uncertainties for *b*-quark jets
JET_EffectiveNP_Detector{1-2}	Energy scale uncertainties due to in-situ calibration
JET_EffectiveNP_Mixed{1-3}	Energy scale uncertainties due to in-situ calibration
JET_EffectiveNP_Modelling{1-4}	Energy scale uncertainties due to in-situ calibration
JET_EffectiveNP_Statistical{1-6}	Energy scale uncertainties due to in-situ calibration
JET_EtaIntercalibration_Modelling	Energy scale uncertainties to cover η-intercalibration non-closure
JET_EtaIntercalibration_NonClosure_highE	Energy scale uncertainties to cover η-intercalibration non-closure
JET_EtaIntercalibration_NonClosure_negEta	Energy scale uncertainties to cover η-intercalibration non-closure
JET_EtaIntercalibration_NonClosure_posEta	Energy scale uncertainties to cover η-intercalibration non-closure
JET_EtaIntercalibration_TotalStat	Energy scale uncertainties to cover η-intercalibration non-closure
JET_FlavComp	Energy scale uncertainty related to flavour composition (response)

(continued)

Table 6.1 (continued)

Systematic uncertainty name	Description
Small-R jets	
JET_Flavor_Response	Energy scale uncertainty related to flavour response
JET_PU_{OffsetMu(NPV),PtTerm,RhoTopology}	Energy scale uncertainties due to pile-up effects
JET_PunchTroughMC16	Energy scale uncertainty for 'punch-through'
JET_SingleParticle_HighPt	Energy scale uncertainty for the behaviour of high-p_T jets
JET_JvtEfficiency	JVT efficiency uncertainty
JER_DataVsMC	Energy resolution uncertainty
JER_EffectiveNP_{1-6,7restTerm}	Energy resolution uncertainties
Large-R jets	
FJ_JMSJES_Baseline	Energy and mass scale uncertainty due to basic data-simulation differences
FJ_JMSJES_Modelling	Energy and mass scale uncertainty due to simulation differences
FJ_JMSJES_Tracking	Energy and mass scale uncertainty on reference tracks
FJ_JMSJES_TotalStat	Energy and mass scale uncertainty from stat. unc. on the measurement
FJ_JER	Energy resolution uncertainty
FJ_JMR	Mass resolution uncertainty
***b*-tagging: VR track jets**	
FT_B_{0-4}	*b*-tagging efficiency uncertainties for *b* jets
FT_C_{0-3}	*b*-tagging efficiency uncertainties for *c* jets
FT_Light_{0-3}	*b*-tagging efficiency uncertainties for light jets
FT_extrapolation	*b*-tagging efficiency uncertainty for high-p_T *b*-quark and *c*-quark jets
FT_extrapolation_from_charm	*b*-tagging efficiency uncertainty on τ-lepton jets
E_T^{miss} and $E_{T,trk}^{miss}$	
MET_SoftTrk_ResoPara(Perp)	'soft term'-related longitudinal(transverse) resolution uncertainty
MET_SoftTrk_Scale	'soft term'-related scale uncertainty
MET_JetTrk_Scale	$E_{T,trk}^{miss}$ scale uncertainty

Luminosity Uncertainty

The luminosity is evaluated by combining different luminosity measurements of several detectors. A precise measurement of the integrated luminosity is a key component of the ATLAS physics program for cross-section measurements in particular because the luminosity measurement is often one of the leading sources of uncertainty. The systematic uncertainty of the luminosity is obtained from the comparison of measurements from different detectors. Considering the full Run 2 data sample, the uncertainty on the luminosity is 2% [1].

Pile-up Uncertainty
In simulated samples only the best-guess of the data pile-up conditions are considered. Once the data are collected, the simulated pile-up conditions are corrected in the simulation to match the condition found in the collected data. This procedure is called *pile-up reweighting* and it is used to correct the distribution of the average number of interactions per bunch crossing in the simulated samples. A systematic uncertainty, as large as the rescaling, is assigned to this procedure and it is of the order of 3%.

Leptons Uncertainties
Systematic uncertainties of electrons and muons are determined for the reconstruction, identification and isolation efficiency as well as the energy scale and resolution (see Sects. 4.2.1 and 4.2.2). In addition, dedicated uncertainties are used to take into account differences in the single electron trigger efficiency. Due to the differences in the reconstruction of electrons and muons, their systematic uncertainties are derived and treated separately. For the muons dedicated systematic uncertainties are also considered for the track-to-vertex association efficiency and for the identification of muons with low transverse momenta ($p_{\mathrm{T}} < 15$ GeV). The lepton uncertainties have a very small impact on the analysis result.

Small-*R* Jet Uncertainty
The systematic uncertainties affecting the small-*R* jet energy calibration are the Jet Energy Scale (JES) and the Jet Energy Resolution (JER) uncertainties. As described in Sect. 4.3.3 the calibration of the energy scale of the small-*R* jets involves several steps, therefore multiple sources of systematic uncertainties are considered. A set of 27 decorrelated uncertainties is chosen for the JES, accounting for the different effects such as η-intercalibration, in-situ calibration, high-p_{T} jets, pile-up, flavour composition and punch-through jets. The total uncertainty on the JES is 4.5% at p_{T} = 20 GeV and it decreases to 1% at p_{T} = 200 GeV.

The total uncertainty applied to the JER is ranging between 10 and 20% for jets with p_{T} = 20 GeV, depending on the η value, and less than 5% for jets with p_{T} > 200 GeV. The JER uncertainties take into account differences between data and simulated events and experimental uncertainties related to the method adopted to measure the JER.

Another source of uncertainty connected to the small-*R* jets is related to the Jet Vertex Tagger efficiency and it is of the order of a few percent.

Large-*R* Jet Uncertainty
The uncertainties that mostly affect the analysis described in this thesis are the one related to the large-*R* jet which is used to reconstruct the Higgs candidate. The total uncertainty of the JES and the Jet Mass Scale (JMS) of the large-*R* jets is of the order of 5%. Four contributions to the JES and JMS uncertainties are implemented in the fit: the *baseline* component accounting for the data-simulation difference, the *modelling* component accounting for the difference among alternative samples, *tracking* component accounting for the reference tracks used in the calibration and the *statistical* component accounting for the statistical uncertainties on the auxiliary

measurement. These uncertainties are treated correlated between the JES and JMS. The systematic uncertainties related to the *baseline* and *modelling* components are treated uncorrelated for all the processes. A relative uncertainty of 20% is assumed for the Jet Mass Resolution (JMR) and the same correlation strategy of the *baseline* and *modelling* components of JMS and JES is adopted.

An additional uncertaintics on the JER of the large-R jet of 2% is considered in the analysis.

b-Tagging Uncertainties

The identification of b-jets is the crucial part of the analysis in order to reconstruct the Higgs boson decay and to suppress background processes containing light-flavour jets. The b-tagging uncertainty arises because scaling factors are applied to simulated events to correct for different tagging efficiencies observed between data and simulations [2]. The uncertainties on these scale factors introduce systematic uncertainties in the analysis. The scale factors and the corresponding systematic uncertainties are derived separately for b-, c- and light-flavour jets. Using eigenvector decompositions, these uncertainties are decomposed into 5, 4, 6 components for b-, c- and light-flavour jets respectively. There is also an additional extrapolation uncertainty for high p_T b-jets and c-jets, and extrapolation uncertainty for τ-lepton jets. The approximate size of the uncertainty on the tagging efficiency is 5% for b-jets, 10% for c-jets and 30% for the light-flavour jets. All the b-tagging uncertainties are treated as uncorrelated for b-jets, c-jets and light-flavour jets in the final fit.

E_T^{miss} Uncertainty

The E_T^{miss} is calculated using physics objects and a soft term. The systematic uncertainties on the calibration and resolution of physics objects are propagated coherently in the E_T^{miss} calculation, while for soft term dedicated uncertainties are considered in the analysis. An additional uncertainty on $E_{T,trk}^{miss}$ is considered for tracks in jets covering data-to-simulation differences. No E_T^{miss} trigger uncertainties are considered since the trigger is fully efficient in the analysis phase space.

6.3 Modelling Uncertainties

The signal and background processes considered in the analysis are mainly modelled using MC generators. The accuracy of the MC samples is a crucial part of the analysis and the uncertainties on their prediction are carefully considered. The composition of the background is different in each of the three lepton channels and regions. All the modelling studies have been done using the Truth Tagging (TT) strategy to mitigate the low statistics of MC samples due to the high light- and c-jet rejections of the b-tagging algorithm. A brief description of the TT strategy can be found in the Sect. 6.3.1.

The modelling uncertainties are derived for all the simulated samples and they cover three effects: cross-section normalisation, relative acceptance in the various analysis regions and large-R jet mass shape. The normalisation uncertainties are taken from the most accurate theory predictions used for the normalisation of the samples. The relative acceptance systematics are derived from the comparison of the nominal sample with the alternative samples, normalized to the same cross-section. The uncertainty is computed according to the double ratio formula:

$$\frac{\text{Yields}_{\text{A}}[\text{alternative}]}{\text{Yields}_{\text{B}}[\text{alternative}]} \Big/ \frac{\text{Yields}_{\text{A}}[\text{nominal}]}{\text{Yields}_{\text{B}}[\text{nominal}]} - 1 \tag{6.1}$$

where Yields_{A} and Yields_{B} are the number of selected events in the region A and B, respectively. Using the double ratio formula, in the $VH(H \to b\bar{b})$ analysis the following uncertainties are considered:

- medium-to-high p_{T}^{V} relative acceptance uncertainties;
- SR-to-CR relative acceptance uncertainties;
- high-to-low purity SR relative acceptance uncertainties;
- flavour composition acceptance[1] uncertainties;
- extrapolation among channels uncertainties.

The alternative samples are either alternative MC samples generated with different MC generators or nominal MC samples with varied parameters implemented as internal weights. The presence of the internal weights allows to easily derive the effects of the systematics variations using the full statistics of the nominal samples. For the relative acceptance uncertainties, the value used in the statistical model described in Chap. 7 is computed summing in quadrature the effects of all the considered variations among the alternative samples.

Uncertainties on the large-R jet mass shape are studied in each analysis region separately comparing the shapes, scaled to have the same normalisation in each region, of the nominal and the alternative samples. The shape uncertainties are derived only for the large-R jet mass m_J variable because it is the discriminating variable used in the final fit.

The following subsections will be dedicated to the signal and background modelling after the description of the truth tagging method. For each process there will be a brief description of the nominal samples followed by the explanation of the systematics uncertainties applied in the fit.

6.3.1 *Truth Tagging Strategy*

The truth tagging strategy is used to reduce the statistical error of simulated dataset using b-tagging techniques. Applying the b-tagging requirements, the amount of selected events is significantly reduced. The approach of applying the b-tagging algorithm is called Direct Tagging (DT). In the analysis the average b-tagging efficiency is 70%. Requiring two b-tagged jets in simulated events, almost half of the

[1] The flavour composition uncertainties is specific for the V + jets samples.

events with two b-labelled jets[2] will be selected. This reduction is even more drastic for c- and light-jets. The idea of the truth tagging is to reweight all the simulated events such that the effect of the tagging algorithm is simulated. For each jet, a weight corresponding to the parametrized efficiency (for b-jets) and the mistag rate (for light- and c-jets) is assigned. The weight depends on the flavour, the $p_{\rm T}$ and η of the jet. All the simulated events are considered, the events with two b-labelled jets have a weight almost equal to 0.49. This algorithm improves the statistic precision of 40% in case of b-jets and even more for light- and c-jets. In this procedure, each jet is treated independently. In the boosted environment the assumption of independence may led to mis-modelling since the tagging performance depends on the close-by hadronic activity. In the $VH(b\bar{b})$ analysis, the truth tagging is only applied to the leading and sub-leading VR track-jets ghost-matched to the Higgs candidate jet.

An alternative approach is the Hybrid Tagging (HT) in which the b-tagging algorithm is applied to the b-jets in the event and the truth tagging to c- and light jets. This approach would remove the potentially dangerous assumption of independence for events with b-labelled jets (90% of the events selected by the analysis), retaining the huge gain in the statistical precision for events with c- and light jets.

Figure 6.1a shows the m_J distribution for the direct, hybrid and truth tagging methods in case of a $t\bar{t}$ sample and relative to the HP SR, medium $p_{\rm T}^V$ region in the 1-lepton channel. A clear bias from the truth tagging is seen in the m_J distribution. The bias is significantly decreased when applying the hybrid approach, indicating that a substantial mis-modelling originates from the assumption of independence.

The differences between the direct tagging and the other two tagging strategies suggest not to use the TT and the HT in the event selection of the analysis. However, a way to exploit the TT is to use it in the modelling studies when ratios between

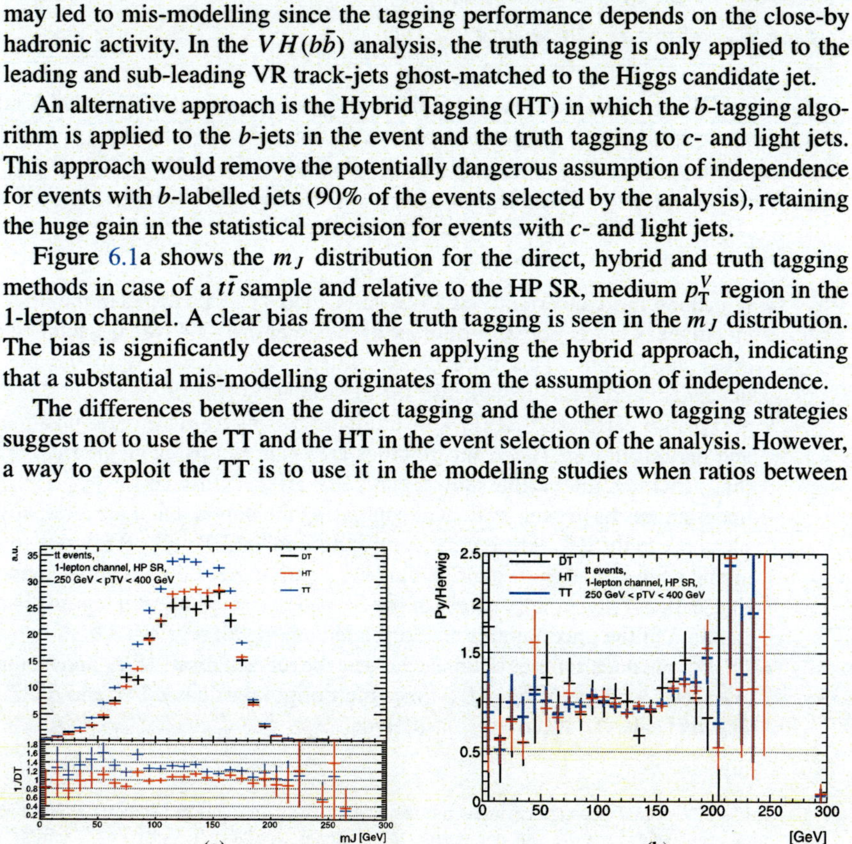

(a) (b)

Fig. 6.1 **a** m_J distributions obtained applying the direct (*black line*), hybrid (*red line*) and truth tagging (*blue line*) using $t\bar{t}$ sample in the HP SR, medium $p_{\rm T}^V$ region in 1-lepton channel. **b** Comparison between $t\bar{t}$ nominal (Poweg + Pythia) and alternative (Poweg + Herwig) samples applying the direct (*black line*), hybrid (*red line*) and truth tagging (*blue line*) in the HP SR, medium $p_{\rm T}^V$ region in 1-lepton channel

[2] The jet labelling is done using the truth information of the generator's event record.

nominal and alternative MC samples are computed. In these ratios, the effect of the bias is cancel out and the statistics of the generators is improved. Figure 6.1b shows the ratio between the nominal and alternative samples for the $t\bar{t}$ sample in the HP SR, medium p_T^V bin in 1-lepton channel. The truth tagging and the hybrid tagging have the same trend and they both agree with the direct tagging distribution. Since with the hybrid tagging the statistics is reduced, the truth tagging is applied for all the modelling studies.

6.3.2 Signal Modelling

The VH signal includes the ZH and WH production modes. The ZH production mode is furthermore split into two contributions: $qq \rightarrow ZH$ and $gg \rightarrow ZH$. The nominal qq-induced samples are generated with `Powheg MiNLO + Pythia 8`, while the gg-induced samples are generated with `Powheg + Pythia 8`.

The VH systematics uncertainties considered in the analysis when performing a signal strength measurement are:

- branching ratio uncertainty of 1.6% and called `TheoryBRbb` in the fit;
- electroweak correction uncertainties of the order of 1–3% for unaccounted higher-order corrections. This uncertainty, called `SysVHNLOEWK` in the fit, is p_T^V-dependent and it can change also the m_J shape.
- acceptance parton shower (PS) uncertainties arising from the hadronization model and parton shower. These uncertainties are evaluated using an alternative sample that generates the parton shower with `Herwig 7` instead of `Pythia 8`. The uncertainties have been evaluated comparing the nominal and the alternative samples only in the SR, splitting the contribution in the HP and LP SR, due to the low signal contamination in the CR. Table 6.2 summarizes the acceptance uncertainties, called `TheoryUEPSAcc` in the fit, and extracted in each lepton channel. In the final fit the uncertainties are correlated among all regions.
- QCD scale uncertainties obtained varying the renormalisation μ_R and factorization μ_F scales by a factor 2 and 0.5 times their original values μ_R^{norm} and μ_F^{norm}. At the end six scale variations are considered:

Table 6.2 Signal acceptance uncertainties arising from the parton shower model. The quoted numbers have been extracted comparing the nominal `Powheg + Pythia 8` sample with the alternative `Powheg + Herwig 7` sample

	0-lepton		1-lepton		2-lepton
	HP SR (%)	LP SR (%)	HP SR (%)	LP SR (%)	SR (%)
Med. p_T^V	+2.4	+4.4	+0.9	+5.2	+3.7
High p_T^V	+0.0	+7.4	+2.8	+6.5	+4.7

Table 6.3 QCD scale migration uncertainties arising from the signal split in p_T^V. The uncertainties are derived from Ref. [3]

Truth p_T^V range (GeV)	Δ_Y^{qq} (%)	Δ_{75}^{qq} (%)	Δ_{150}^{qq} (%)	Δ_{250}^{qq} (%)	Δ_{400}^{qq} (%)	Δ_Y^{gg} (%)	Δ_{75}^{gg} (%)	Δ_{150}^{gg} (%)	Δ_{250}^{gg} (%)	Δ_{400}^{gg} (%)
[150, 250[	0.7%	3.3	1.3	−0.41	–	25	26	13	−2.6	–
[250, 400[	0.7	3.3	1.3	1.4	−0.38	25	26	13	14	−1.3
[400, ∞[	0.7	3.3	1.3	1.4	1.8	25	26	13	14	15

$$\left[\frac{\mu_R}{\mu_R^{norm}}, \frac{\mu_F}{\mu_F^{norm}}\right] : [0.5, 1][1, 0.5][2, 1][1, 2][0.5, 0.5][2, 2] \tag{6.2}$$

The QCD scale uncertainties have been rearranged into seven contributions:

- uncertainty on the overall cross-section (Δ_Y);
- uncertainties for the migration between different p_T^V bins (Δ_{75}, Δ_{150}, Δ_{250}, Δ_{400});
- uncertainties for the jet migration (Δ_1, Δ_2);

All the QCD acceptance uncertainties, called `QCDScaleDelta` in the fit, are calculated separately for $qq-$ and $gg-$initiated signal samples and using the truth p_T^V information. Tables 6.3 and 6.4 summarize their values.

- Parton Distribution Function (PDF) + α_S uncertainties evaluated varying the PDF or the QCD coupling constant α_S. The PDF dependence is evaluated by comparing the nominal sample with 30 alternative samples with varied PDF functions. In addition there is also the α_S variation evaluated with respect to the nominal sample. The uncertainties are calculated separately for WH, $qq \to ZH$ and $gg \to ZH$ processes in each analysis region. The PDF uncertainties called

Table 6.4 QCD scale migration uncertainties arising from the signal split in the number of additional jets. The uncertainties are derived from Ref. [3]

Truth p_T^V range (GeV)	Δ_1^{qq}			Δ_2^{qq}		
	0 jet (%)	1 jet (%)	≥2 jets (%)	0 jet (%)	1 jet (%)	≥2 jets (%)
[150, 250[	−4.1	5.1%	5.1	–	−5.0	8
[250, 400[	−5.4	5.3	5.3	–	−5.0	6.7
[400, ∞[	−6.8	5.5	5.5	–	−5.7	6.7
Truth p_T^V range (GeV)	Δ_1^{gg}			Δ_2^{gg}		
	0 jet (%)	1 jet (%)	≥2 jets (%)	0 jet (%)	1 jet (%)	≥2 jets (%)
[150, 250[	−50	26	26	–	−20	26
[250, 400[	−100	28	28	–	−38	28
[400, ∞[	−100	30	30	–	−66	30

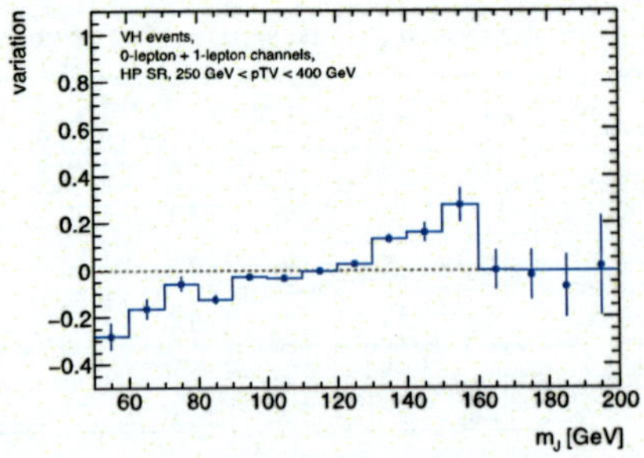

(a) 0+1L, HP SR, med. p_{T}^{V}

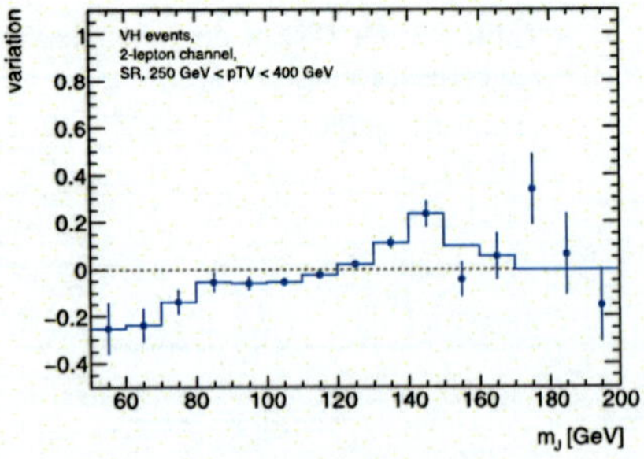

(b) 2L, SR, med. p_{T}^{V}

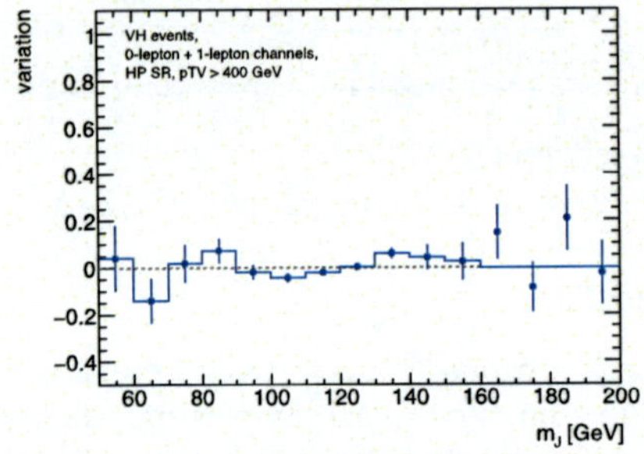

(c) 0+1L, HP SR, high p_{T}^{V}

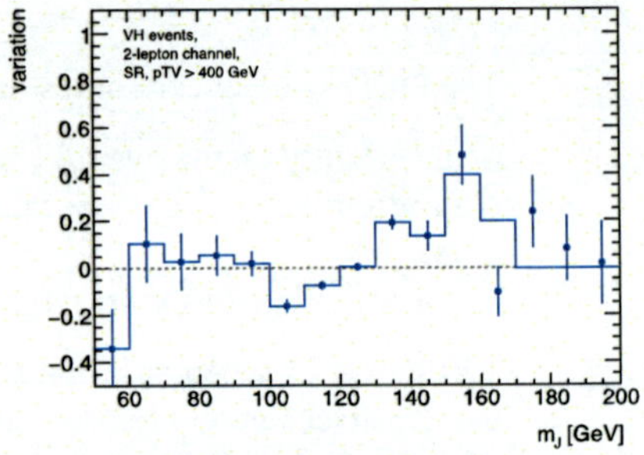

(d) 2L, SR, high p_{T}^{V}

Fig. 6.2 m_J variation due to the change of PS generator from `Pythia 8` to `Herwig 7`. The contributions in the 0- and 1-lepton channels are summed. The variation `Herwig 7` w.r.t. `Pythia 8` is depicted using points whereas the final fit template is shown as a *solid blue line*

`TheoryPDF_{1-30}` are of the order of 1–2 and 2–5% for the $qqVH$ and $ggZH$ processes, respectively. The α_S uncertainties called `Theoryalphas` are of the order of 1–2 and 1–3% for $qqVH$ and $ggZH$, respectively.

In addition to the PS acceptance uncertainties, a PS shape uncertainties is considered. The shape uncertainties from the PDF variations and QCD scales are negligible. The PS shape uncertainty is evaluated comparing the nominal sample with the alternative sample in which `PS` is simulated using the `Herwig 7` generator instead of `Pythia 8`. The m_J shape uncertainty, called `TheoryUEPSShape` in the fit, has been derived individually in the medium and high p_{T}^{V} regions and in each lepton channels. Due to similar shape variation in the HP SR of the 0- and 1-lepton channels, the two contributions are assumed identical. As for the acceptance PS uncertainties, in the final fit the shape uncertainties are kept correlated between all regions. Figure 6.2 shows the m_J shape variation implemented in the final fit.

Table 6.5 shows a summary of the signal uncertainties applied in the analysis when performing a signal strength measurement.

Performing a cross-section measurement, the overall uncertainties on the cross-section are removed because they are moved to the theory predictions. Nevertheless

Table 6.5 Systematic uncertainties estimated for the signal simulation. The first column states the name of the uncertainty used in the fit, followed by a brief description in the second column. The third column lists the analysis region in which the uncertainty is applied, the fourth the value of the uncertainty and the last column states the effect of the uncertainty

Name	Description	Samples/categories	Value	Effect
TheoryBRbb	BR variation	$VHbb$, all regions	1.6%	Normalisation
QCDScaleDeltaY	QCD scales variation (σ)	$qqVH$, $ggZH$, all regions	0.7% ($qqVH$), 25% ($ggZH$)	Normalisation
QCDScaleDelta75,150,250,400	QCD scales variation (p_T^V)	$qqVH$, $ggZH$, all regions	1–3% ($qqVH$), 10–20% ($ggZH$)	Normalisation
QCDScaleDelta1,2	QCD scales variation (n_{jet})	$qqVH$, $ggZH$, all regions	4–7% ($qqVH$), 20–100% ($ggZH$)	Normalisation
TheoryUEPSAcc	PS variation	$VHbb$, all regions	1–7%	Normalisation
TheoryUEPSShape	PS variation	$VHbb$, all regions	–	Shape
TheoryPDF_{1-30}	PDF variations	$qqVH$, $ggZH$, all regions	1–2% ($qqVH$), 2–5% ($ggZH$)	Normalisation
Theoryalphas	α_S variation	$qqVH$, $ggZH$, all regions	1–2% ($qqVH$), 1–3% ($ggZH$)	Normalisation
VHNLOEWK	NLO EW m_J var.	$VHbb$, all regions	–	Shape + normalisation

residual uncertainties need to be considered. They have been evaluated using the *maximum splitting* categorization and merged to the binning scheme chosen for the measurement. The $VH(b\bar{b})$ *boosted* analysis performs cross-section measurements in four bins:

- WH, $p_T^{t,V} \in [250, 400[$ GeV;
- WH, $p_T^{t,V} \in [400, \infty[$ GeV;
- ZH, $p_T^{t,V} \in [250, 400[$ GeV;
- ZH, $p_T^{t,V} \in [400, \infty[$GeV;

Calling δ_t the systematic uncertainty on the STXS bin t of the *maximum splitting*, if the cross-section is performed on a coarser bin g that contains a set of bins t, the residual uncertainty $\delta_{t,residual}$ is evaluated subtracting the overall uncertainty δ_g on bin g from δ_t:

$$\delta_{t,residual} = \delta_t - \delta_g \tag{6.3}$$

The overall uncertainty δ_g on the coarser bin g is given by:

$$\delta_g = \frac{\sum\limits_{t \in group} \delta_t \sigma_t}{\sum\limits_{t \in group} \sigma_t} \tag{6.4}$$

where σ_t is the cross-section of the bin t.

In the analysis, residual uncertainties for the PDF + α_S and QCD scales replace the PDF + α_S and QCD scale uncertainties, respectively. The residual uncertainties are evaluated for each component of the PDF, α_S and QCD scale uncertainties.

6.3.3 Top Pair Production Modelling

The top quark pair production is one of the main backgrounds in 0- and 1-lepton channels. To better constrain and study this process, top enriched CRs are defined by requiring additional b-tagged track-jets outside the Higgs candidate jet. The $t\bar{t}$ nominal sample is generated with `Powheg + Pythia 8`.

In 0- and 1-lepton channels the leading decay mode of the $t\bar{t}$ pair is the semi-leptonic one in which one W boson decays hadronically and the other one decays leptonically. Figure 6.3 shows the large-R jet mass distribution m_J in 0- and 1-lepton channels in the SR[3] and in the CR. In each analysis region, the shapes obtained in each lepton channel are consistent in both SR and CR. The large-R jet mass of $t\bar{t}$ events is smaller than the top quark mass due to a non-complete containment of the top decay inside the large-R jet. In fact, the hadronic decay of the W boson together with one b-quark can be partially or fully reconstructed within the large-R jet. These topologies can be recognised in Fig. 6.3 where the shapes show a peaking structure due to events with two VR track-jets or events with three or more VR track-jets inside the large-R jet. The ratio of the two components strongly depends on p_{T}^{V}. In the medium p_{T}^{V} region, in both SR and CR almost half of the events have two VR track-jets inside the large-R jet, while in the high p_{T}^{V} region up to 70% of the events have three or more VR track-jets inside the large-R jet. The large-R jet mass m_J distribution of events with two VR track-jets in the large-R peaks at ~120 GeV (the top quark is partially reconstructed in the large-R jet) while the distribution of events with at least three VR track-jets inside the large-R peaks at ~175 GeV (the top is fully reconstructed in the large-R jet). The difference between the events with two and three or more VR track-jets inside the large-R jet causes the introduction of a systematic uncertainty, as described later.

Furthermore, the flavour composition of the large-R jet is studied. In both 0- and 1-lepton channels, the two leading track-jets are b-labelled and a c-labelled jets in 75% of the $t\bar{t}$ events.

Several alternative $t\bar{t}$ samples have been introduced with the goal of assessing modelling systematics uncertainties for this process. The sources of systematics uncertainties are:

- Parton Showers (PS) uncertainties evaluated with the alternative sample in which the parton shower is simulated using `Herwig 7` instead of `Pythia 8` MC generator.

[3] In the following "SR" indicates the inclusive SR which is the result of the merging of the HP SR and LP SR.

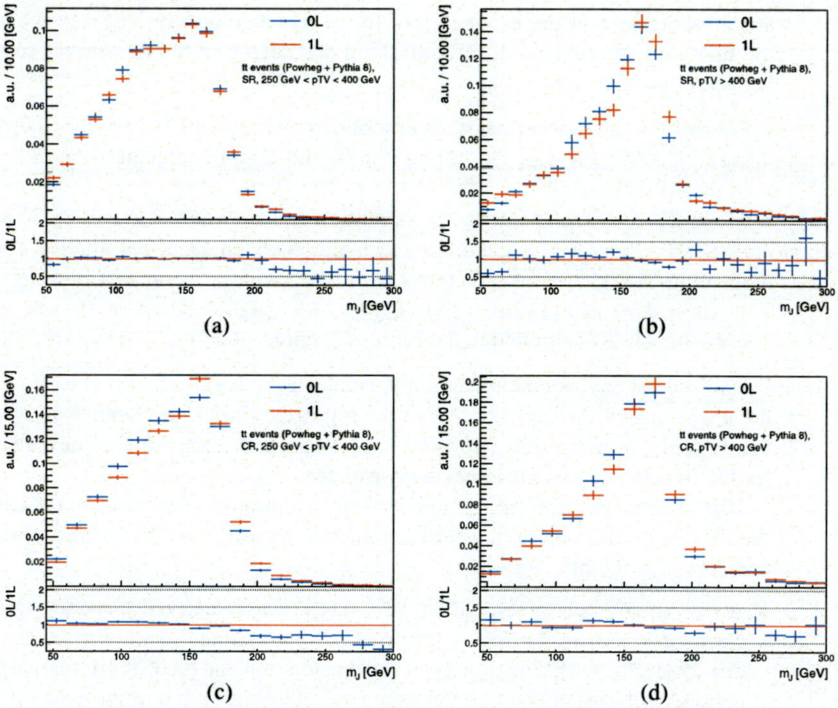

Fig. 6.3 Normalised large-R jet mass m_J distribution from $t\bar{t}$ process in the SR (*top row*) and CR (*bottom row*) in medium p_{T}^{V} (*left side*) and high p_{T}^{V} (*right side*) region in 0-lepton (*blue line*) and 1-lepton (*red line*) channels

- Matrix Element (ME) uncertainties evaluated with the alternative sample in which the matrix element calculation is simulated using `MadGraph 5` instead of `Powheg` MC generator.
- Initial State Radiation (ISR) uncertainties evaluated using internal weight variations in the nominal sample.
- Final State Radiation (FSR) uncertainties evaluated using internal weight variations in the nominal sample.

The $t\bar{t}$ process is one of the main background in the analysis so its normalisation is a free floating parameter in the fit. Consequently no normalisation uncertainties are extracted for the $t\bar{t}$ process in 0- and 1-lepton channels. Two separate normalisation factors, called `norm_ttbar_L0` and `norm_ttbar_L1`, are used for $t\bar{t}$ events in 0- and 1-lepton channels because separate dedicated control regions are available. The $t\bar{t}$ background in the 2-lepton channel is negligible so an overall normalisation uncertainty, called `ttbarNorm_L2`, of 20% is applied in the 2-lepton channel.

The most significant source of acceptance uncertainty is from the matrix element calculation, followed by the parton shower uncertainties. FSR and ISR uncertainties are sub-dominant in the evaluation of the acceptance uncertainties. In the analysis the following acceptance uncertainties are considered in the fit with no correlation between 0- and 1-lepton channels:

- HP-to-LP relative acceptance uncertainties, called `R_HPLP_SR_L0_J0_Ttbar` and `R_HPLP_SR_L1_J0_Ttbar` in 0- and 1-lepton channels, of 18% applied in the HP SR;
- SR-to-CR relative acceptance uncertainties, called `R_SRCR_SR_L0_Ttbar` and `R_SRCR_SR_L1_Ttbar` in 0- and 1-lepton channels, of 6% applied in the SR;
- medium-to-high p_{T}^{V} relative acceptance uncertainties, called `R_Ptv_400_L0_Ttbar` and `R_Ptv_400_L1_Ttbar` in 0- and 1-lepton channels, of 20% applied in the high p_{T}^{V} region[4];

In addition an acceptance uncertainty, called `R_ttbarVRjets_ttbar2VR`, for the modelling of the containment of the top decay inside the large-R jet of 20% is applied only on events with two VR track-jets inside the large-R jet. The uncertainties for the 0- and 1-lepton channels are correlated.

The m_J shapes uncertainties are derived by summing the shapes in the CR and in the SR due to the consistency between the two region. Two shapes uncertainties are considered in the fit:

- A m_J shape uncertainty, called `TTbarMJISR`, is derived considering the ISR variation where a tail effect of the order of 20% in both 0- and 1-lepton channels is observed (Fig. 6.4). The ratio between the nominal and the ISR alternative samples is fitted in 1-lepton channel through a second order polynomial function (yellow line in Fig. 6.4) and then propagated to the global fit in 0- and 1-lepton channels. One shape uncertainty is derived in the medium p_{T}^{V} region and another in the high p_{T}^{V} region.
- An other m_J shape uncertainty, called `TTbarMJPS`, is given by the parton shower uncertainties. The comparison between the nominal and the parton shower alternative samples shows an oscillating behaviour around the top quark mass in both 0- and 1-lepton channel. In this case the ratio of the nominal over the alternative samples (bottom panel of Fig. 6.5) is propagated to the global fit without any fit on the ratio itself. The shape propagated to the fit corresponds to the sum of 0- and 1-lepton channels histograms, inclusively in both p_{T}^{V} regions and, SRs and CRs.

Table 6.6 shows a summary of all the $t\bar{t}$ modelling systematics uncertainties.

6.3.4 Single-Top Production Modelling

The single-top production is a background in 0- and 1-lepton channels. This background is sub-dominant with respect to the $t\bar{t}$ pair production but in some of the SRs,

[4] Comparing nominal and alternative samples the difference is less than 20% but an higher value is set due to a mis-modelling measured in the mass distribution around the top peak.

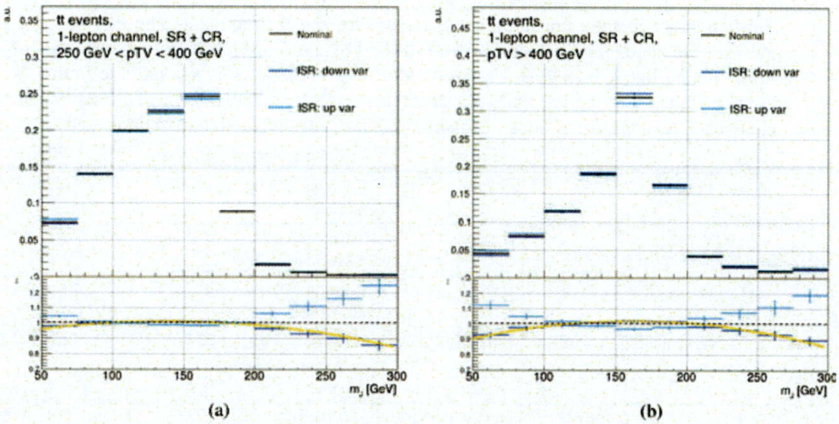

Fig. 6.4 Normalized large-R jet mass m_J distributions for ISR uncertainties for $t\bar{t}$ events in the 1-lepton channel in the 250 GeV $\leq p_T^V < 400$ GeV (**a**) and the $p_T^V \geq 400$ GeV (**b**) regions. The distribution is obtained summing the contribution in the SR and in the CR. The second order polynomial used as fitting function is drawn as a *dashed yellow line*

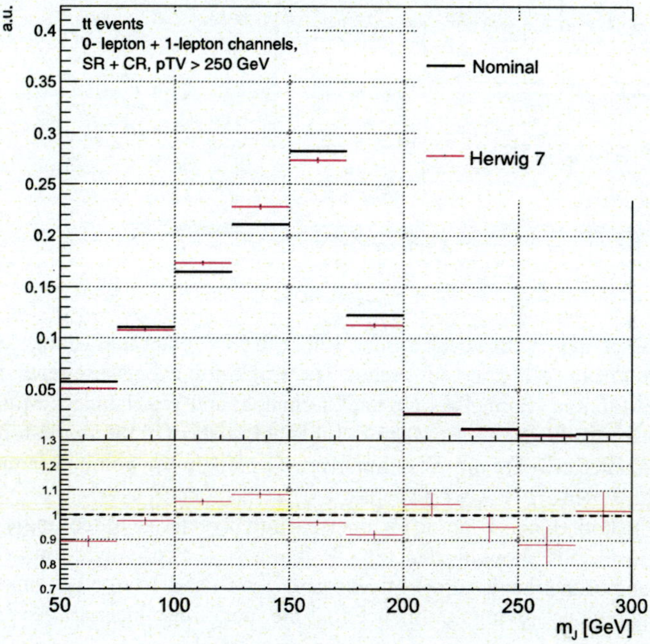

Fig. 6.5 Normalized large-R jet mass m_J distributions for nominal (`Powheg + Pythia 8`) and alternative (`Herwig 7`) samples for $t\bar{t}$ events summing the contribution of 0- and 1-lepton channels, SR and CR, and medium and high p_T^V regions

Table 6.6 Systematic uncertainties estimated for the $t\bar{t}$ simulation. The first column states the name of the uncertainty used in the fit, followed by a brief description in the second column. The third column lists the analysis region in which the uncertainty is applied, the fourth the value of the uncertainty and the last column states the effect of the uncertainty. The 0-lepton and 1-lepton channels have separate floating normalisations for $t\bar{t}$ as they both have a dedicated CR

Name	Description	Samples/categories	Value	Effect
`norm_ttbar_L0`	$t\bar{t}$ norm.	0-lepton channel, all regions	Floating	Normalisation
`norm_ttbar_L1`	$t\bar{t}$ norm.	1-lepton channel, all regions	Floating	Normalisation
`ttbarNorm_L2`	$t\bar{t}$ norm.	2-lepton channel, all regions	20%	Normalisation
`R_ttbarVRjets_ttbar2VR`	2VR/3pVR ratio	$t\bar{t}$ (2VR), applied to 0-lepton and 1-lepton	20%	Normalisation
`R_Ptv_400_L0_Ttbar`	Med. to high p_T^V ratio	$t\bar{t}$, applied to 0-lepton, high p_T^V	20%	Normalisation
`R_Ptv_400_L1_Ttbar`	Med. to high p_T^V ratio	$t\bar{t}$, applied to 1-lepton, high p_T^V	20%	Normalisation
`R_HPLP_SR_L0_J0_Ttbar`	HP to LP ratio	$t\bar{t}$, applied to 0-lepton, HP region	18%	Normalisation
`R_HPLP_SR_L1_J0_Ttbar`	HP to LP ratio	$t\bar{t}$, applied to 1-lepton, HP region	18%	Normalisation
`R_SRCR_SR_L0_Ttbar`	SR to CR ratio	$t\bar{t}$, applied to 0-lepton, SR	6%	Normalisation
`R_SRCR_SR_L0_Ttbar`	SR to CR ratio	$t\bar{t}$, applied to 1-lepton, SR	6%	Normalisation
`TTbarMJISR`	ISR m_J shape var.	$t\bar{t}$, all regions	–	Shape
`TTbarMJPS`	PS m_J shape var.	$t\bar{t}$, all regions	–	Shape

it has a shape with a peak around 120–140 GeV, potentially dangerous for a measurement of the Higgs boson events. The single-top events are generated separately for the different channels (s-channel, t-channel and Wt-channel). Mostly all the single-top events (95%) are from the Wt-channel, so most of the single-top uncertainties are evaluated only for the Wt component. The single-top nominal samples are generated with `Powheg + Pythia 8`.

In both 0- and 1-lepton channels, a high percentage of the single-top events (80%) is in the SR showing that the CR is not suited for single-top events studies. The shape of the single-top Wt is consistent in 0- and 1-lepton channels in all the analysis regions, giving confidence that the modelling uncertainties can be treated in a correlated way between the two lepton channels.

In the single-top Wt events the hadronic decay of the W boson, together with the b-quark from the top decay is fully or partially reconstructed within the large-R jet as in the case of the $t\bar{t}$ events. The dominant decay mode of the single-top Wt channel is the semi-leptonic decay in which one W boson decays leptonically and the other

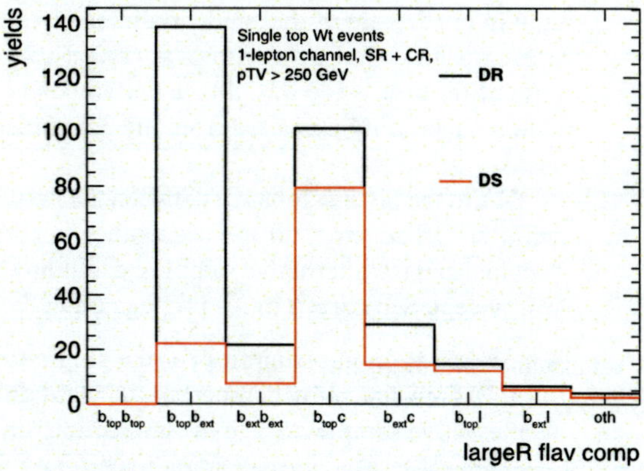

Fig. 6.6 Flavour composition of the large-R jet in 1-lepton channel for nominal (DR scheme) and alternative (DS scheme) sample. The plot is done by summing the contribution of the SR and CR, and the two p_{T}^{V} regions. In the plot the b-quark from the top is indicated as $\mathrm{b_{top}}$, while the spectator b as $\mathrm{b_{ext}}$

W boson decays hadronically. One of these two W boson is produced from the top decay.

The flavour composition of the large-R jet for single-top events in the Wt channel shows that 60–65% of the events in the SR have two b-labelled jets inside the large-R jet, while in the CR 50–60% of the events have a b-labelled and a c-labelled jet, that originate from the W boson and top quark decays. The 0- and 1-lepton channels have the same flavour composition. Due to the same flavour composition and similar shape in 0- and 1- lepton channels, all the uncertainties of the single-top samples are studied in 1-lepton channel and then propagated to the 0-lepton channel. Therefore correlated uncertainties are applied in the two lepton channels.

Alternative single-top Wt samples are used to study the modelling systematics uncertainties for this process. The sources of systematics uncertainties are the same as the one of the $t\bar{t}$ process (PS, ME, ISR, FSR) with an additional contribution that takes into account the single-top -$t\bar{t}$ interference. The nominal single-top Wt sample uses the so-called "diagram removal" (DR) scheme to keep the interference between the two samples while the alternative sample uses the "diagram subtraction" (DS) scheme in which the interface is removed [4]. The difference between the nominal sample and the alternative sample with the DS scheme is called DS-DR variation.

The alternative sample which uses the DS scheme has been investigated in a more detailed way because it gives large differences with respect to the nominal sample. The main difference between the two samples comes from cases where the large-R jet contains the spectator b-quark[5] (Fig. 6.6). In the nominal sample, most of the events have a large-R jet which contains the b-quark from the top decay and the spectator b, while in the alternative sample the large-R jet mainly contains the b- and the c-quark from the top decay.

The theoretical uncertainties affecting the global normalisation are derived for each single-top process. The uncertainties account for the renormalisation and factorisation scales, the α_S uncertainty and the error on the parton density function. The

[5] The "spectator b" is the b-quark not coming from the top decay.

sum in quadrature of these uncertainties gives the total uncertainties of 4.4, 4.6 and 6.2% for the t- (stoptNorm), s- (stopsNorm) and Wt channel respectively. In case of the single-top Wt, this uncertainties is negligible compared to the normalisation uncertainty extracted from the comparison of the nominal and the alternative yields.

Due to the big discrepancy between the nominal sample and the alternative sample with the DS scheme, all the acceptance uncertainties are evaluated comparing the nominal and the alternative samples, excluding the contribution from the DS scheme. The acceptance uncertainties propagated in the analysis are:

- normalisation uncertainties for the single-top Wt sample, called stopWtNorm, of 20% evaluated by comparing the yield difference between the nominal and the alternative samples. Since the uncertainty on the theoretical cross-section is much smaller than this normalisation uncertainty, its contribution is neglected.
- HP-to-LP relative acceptance uncertainty, called R_HPLP_SR_J0_Stop, of 25% applied in the HP SR.
- SR-to-CR relative acceptance uncertainty, called R_SRCR_CR_Stop, of 30% applied in the CR;
- medium-to-high p_T^V relative acceptance uncertainty, called R_Ptv_400_Stop, of 20% applied in the high p_T^V region

Additionally a 0-lepton-to-1-lepton acceptance uncertainty, called StopNorm_L0, of 20% is applied in the 0-lepton channel because all the uncertainties are extrapolated in 1-lepton channel. This uncertainty is evaluated considering also the DS variation which gives the main contribution.

The shape uncertainties are estimated via comparisons of the shape distributions between the nominal and all the alternative samples. The major contribution is from the alternative sample with the DS scheme and all the other contributions can be neglected. The comparison of the nominal sample and the alternative sample with the DS scheme (Fig.6.7) is used to extract the shape uncertainty by summing the contribution of the SR and CR in each p_T^V region. Since all the acceptance uncertainties do not take into account the DS-DR variation, each shape is multiplied by a normalisation factor given by the yield ratio of the nominal sample and the alternative sample with the DS scheme, evaluated separately in 0- and 1-lepton channel. In the fit the DS-DR systematic uncertainty, called StopWtDSDR, is implemented as shape and normalisation systematics.

Table 6.7 shows a summary of all the single-top modelling systematics uncertainties.

6.3.5 Vector Boson + Jets Production Modelling

The production of a single vector boson V in association with jets is one of the main backgrounds in all the lepton channels considered in the analysis. The V+jets

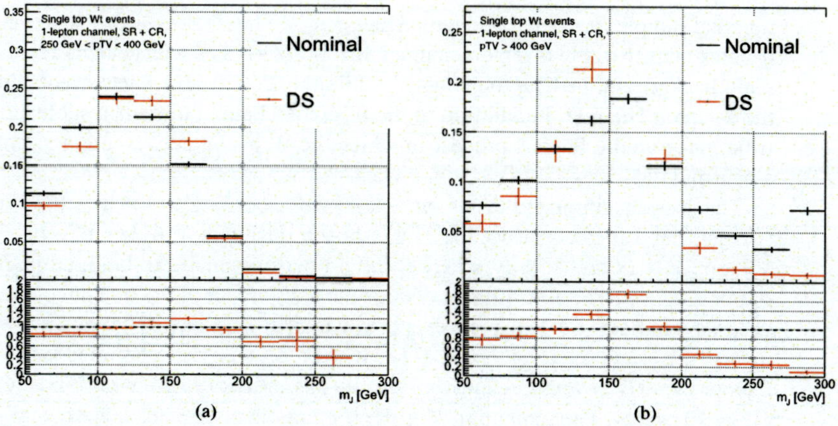

Fig. 6.7 Normalised m_J distributions of the nominal sample and the alternative sample with the DS scheme for the single-top Wt events in the medium (**a**) and high (**b**) p_T^V regions in 1-lepton channel summing the contribution of the SRs and CRs

Table 6.7 Systematic uncertainties estimated for the single-top simulation. The first column states the name of the uncertainty used in the fit, followed by a brief description in the second column. The third column lists the analysis region in which the uncertainty is applied, the fourth the value of the uncertainty and the last column states the effect of the uncertainty

Name	Description	Categories	Value (%)	Effect
`stopWtNorm`	Single-top (Wt-chan.) norm.	All channels, all regions	20.0	Normalisation
`stopsNorm`	Single-top (s-chan.) norm.	All channels, all regions	4.6	Normalisation
`stoptNorm`	Single-top (t-chan.) norm.	All channels, all regions	4.4	Normalisation
`StopNorm_L0`	0-lepton to 1-lepton ratio (from DS/DR)	All channels, applied in 0-lepton	20	Normalisation
`R_Ptv_400_Stop`	Med. to high p_T^V ratio	Single-top, 0-lepton and 1-lepton chan., high p_T^V	20	Normalisation
`R_HPLP_SR_J0_Stop`	HP to LP ratio	Single-top, 0-lepton and 1-lepton chan., HP region	25	Normalisation
`R_SRCR_CR_Stop`	SR to CR ratio	Single-top, applied to 0-lepton and 1-lepton, CR	30	Normalisation
`StopWtDSDR`	DS/DR m_J var.	Wt-chan., all regions	–	Shape + normalisation

nominal samples are generated with `Sherpa 2`. The $Z(Z \rightarrow \nu\bar{\nu})$ + jets events contributes mainly in the 0-lepton channel, the $W(W \rightarrow l\nu)$ + jets events represent the main background in 1-lepton channel while the $Z(Z \rightarrow ll)$ + jets events dominate the 2-lepton channel. In addition in the 0-lepton channel there is a subleading contribution from the W+jets process in which the W boson decays into a τ-lepton and a neutrino.

The V+jets simulated events are classified according to the true flavour of the two jets used to reconstruct the Higgs boson decay as $V + bb$, $V + bc$, $V + bl$. $V + cc$, $V + cl$ and $V + ll$, where l = light parton which can be light quarks as u, d, s or gluons g. The algorithm to deduce the assignment of the true flavour of a jet is described in Sect. 4.3.7. V+jets events in which at least one of the two jets is a b-jet or both are c-jets (i.e. $V + bb$, $V + bc$, $V + bl$. $V + cc$) are classified as V+ heavy flavour (V+HF) events. Almost all the V+jets events (95%) are V+heavy flavour (V+HF) events. The remaining $V + cl/ll$ components are considered negligible so the modelling uncertainties are studied only for the V+HF events.

The large-R jet mass distribution of Z+HF events has a similar shape in the SRs and in CRs in all the lepton channels. Also the W+HF events have a similar shape in the SRs and CRs. The shape of the mass distributions of Z+HF events is consistent between the 0- and 2-lepton channels (Fig. 6.8a). Figure 6.8b shows that the shapes of the mass distributions of W+HF events are also consistent between the 0- and 1-lepton channels. For this reason the uncertainties of W+HF (Z+HF) events are determined in 1-lepton (2-lepton) channel, and then extrapolated to the 0-lepton channel.

The main sources of systematic uncertainties for V+jets events come from the accurancy of the ME predictions and from the parton shower modelling. Two actions are taken to estimate such uncertainties:

- variation of the renormalisation scale μ_R by a factor 2 or 0.5 with respect to its original value μ_R^{norm} in the ME calculation. Such variation is implemented as internal weight in the nominal sample.

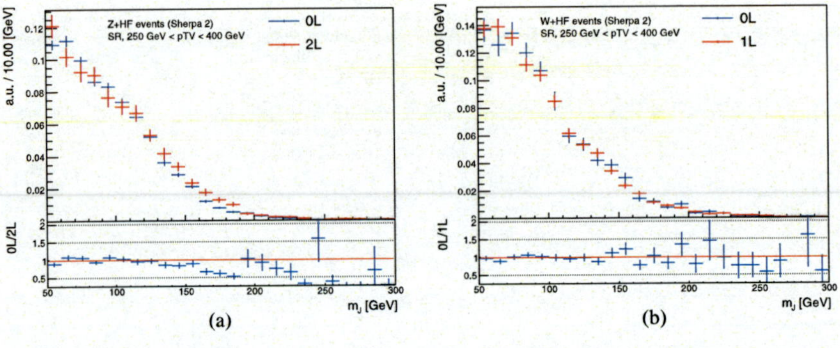

Fig. 6.8 Normalised large-R jet mass m_J distribution of Z+HF (**a**) and W+HF (**b**) in the SR in the different lepton channels

- alternative sample generated using MadGraph for the ME calculation and Pythia for the PS description instead of Sherpa. The comparison between the nominal sample and the alternative sample allows for analysing several effects of mis-modelling in the generation at the same time.

Since the V+HF is one of the main background, floating normalisations, called norm_Zbb and norm_Wbb, are adopted for the Z+HF and W+HF processes. The normalisation factors are correlated across all the analysis regions. For $Z + cl/ll$ and $W + cl/ll$ events the normalisation uncertainties, called WclNorm, WllNorm, ZclNorm and ZllNorm, of 30% are applied in all the regions.

The relative acceptance uncertainties propagated in the analysis are:

- HP-to-LP relative acceptance uncertainty, called R_HPLP_SR_J0_Whf, of 18% for W+HF events applied in the HP SR in both 0- and 1-lepton channels.
- HP-to-LP relative acceptance uncertainty, called R_HPLP_SR_J0_L0_Zhf of 18% for Z+HF events applied in the HP SR in 0-lepton channel.
- medium-to-high p_{T}^{V} relative acceptance uncertainty of 30% for W+HF events in 0-lepton channel, called R_Ptv0L_400_Whf, and of 10% in 1-lepton channel. In the 1-lepton channel the p_{T}^{V} uncertainties are decorrelated in HP and LP SR because the comparison between the nominal and the alternative samples shows an opposite trend in the two regions. The uncertainties in 1-lepton channel are called R_Ptv1L_400_J0_Whf and R_Ptv1L_400_J1_Whf for the HP and LP SR, respectively.
- medium-to-high p_{T}^{V} relative acceptance uncertainty, called R_Ptv_400_Zhf, of 10% for Z+HF events in 0- and 2-lepton channels.
- SR-to-CR relative acceptance uncertainty, called R_SRCR_CR_Whf, of 90% for W+HF events applied in the CR in 0-lepton channel and of 40% in 1-lepton channel. In the 0-lepton channel the big uncertainty is due to the large difference between the nominal and the alternative samples generated with MadGraph+ Pythia. The SR-to-CR uncertainties are treated correlated in the two lepton channels.
- SR-to-CR relative acceptance uncertainty, called R_SRCR_CR_Zhf, of 40% for Z+HF events applied in the CR in 0-lepton channel.
- channel extrapolation uncertainty, called WbbNorm_L0, of 20% for W+HF events applied in 0-lepton channel because modelling studies are performed in 1-lepton channel and then propagated to 0-lepton channel.
- channel extrapolation uncertainty, called ZbbNorm_L0, of 16% for Z+HF events in 0-lepton channel because modelling studies are performed in 2-lepton channel and then propagated to 0-lepton channel.
- flavour composition acceptance uncertainty for the different component (bb, bc, bl, cc) of the V+HF background. An uncertainty of 30% is applied on the bc (ZbcZbbRatio, WbcWbbRatio) and bl (ZblZbbRatio, WblWbbRatio) components, while an uncertainty of 20% is applied on the cc component (ZccZbbRatio, WccWbbRatio).

Shape uncertainties for Z+HF and W+HF events are extracted comparing the nominal sample and the alternative sample with the varied μ_R (Fig. 6.9). The ratio

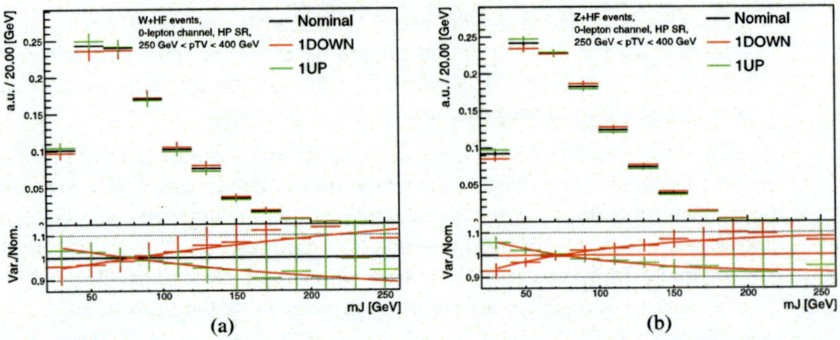

Fig. 6.9 Normalised large-R jet mass distribution of nominal sample and alternative sample with varied μ_R for W+HF (**a**) and Z + HF (**b**) processes in the HP SR, medium p_{T}^{V} region of the 0-lepton channel. The renormalisation scale μ_R has been varied giving an up and down variation

between the nominal and the variation is fitted with an exponential function and a constant, $e^{p_0+p_1x} + p_2$. The parameters of the function are extracted in each analysis regions. The shape uncertainties for W+HF, called `WmJShape`, and Z+HF, called `ZmJShape` events are fully correlated in all the analysis regions in the fit.

A summary of the aforementioned W+jets and Z+jets uncertainties is presented in Tables 6.8 and 6.9, respectively.

6.3.6 Diboson Production Modelling

The diboson background consists of final states arising from WZ, ZZ and WW events. The diboson process is an important background in all the lepton channels because the large-R jet mass distribution has a peaking structure near the Higgs boson mass. When the Z boson decays into two b-quarks, the final state of the VZ process is similar to the signal process. Even if the diboson process is a subdominant contribution, it provides an important reference for the validation. For this reason the VZ process is measured simultaneously with the VH signal in the final fit.

The diboson nominal samples are generated with `Sherpa 2`. The main contribution from the diboson sample is coming from WZ and ZZ events, the contribution of WW events is very small (1–2%). Due to the low yields, modelling uncertainties for WW events are not derived.

Figure 6.10 shows the m_J distribution of WZ and ZZ events in the SR with $p_{\mathrm{T}}^{V} > 250$ GeV in 0- and 1-lepton channels. The shapes are consistent between the lepton channels. Moreover, in each lepton channel a similar shape has been found in the different analysis regions (HP SR, LP SR and CR). To reduce the statistical fluctuations, the shape uncertainties are extracted summing the contribution of the different channels and regions.

Table 6.8 Systematic uncertainties applied to the W + jets simulation. The first column states the name of the uncertainty followed by a brief description in the second column. The third column lists the samples the uncertainty is applied to and in which analysis regions, the fourth the value of uncertainty and the last column states the effect of the uncertainty

Name	Description	Samples/categories	Value	Effect
`norm_Wbb`	*Whf* norm.	*Whf*, applied to all channels, all regions	Floating	Normalisation
`WbbNorm_L0`	*Whf* norm. ratio	*Whf*, 0-lepton to 1-lepton ratio, applied to 0-lepton, all regions	20%	Normalisation
`WclNorm`	*Wcl* norm.	*Wcl*, all regions	30%	Normalisation
`WlNorm`	*Wl* norm.	*Wl*, all regions	30%	normalisation
`WbcWbbRatio`	*Wbc/Wbb* ratio	*Wbc*, applied to 0-lepton and 1-lepton	30%	Normalisation
`WblWbbRatio`	*Wbl/Wbb* ratio	*Wbl*, applied to 0-lepton and 1-lepton	30%	Normalisation
`WccWbbRatio`	*Wcc/Wbb* ratio	*Wcc*, applied to 0-lepton and 1-lepton	20%	Normalisation
`R_Ptv0L_400_J0_Whf`	med. to high p_T^V ratio	*Whf*, applied to 0-lepton, high p_T^V	30%	Normalisation
`R_Ptv1L_400_J0_Whf`	med. to high p_T^V ratio	*Whf*, applied to 1-lepton, high p_T^V, HP SR	10%	Normalisation
`R_Ptv1L_400_J1_Whf`	med. to high p_T^V ratio	*Whf*, applied to 1-lepton, high p_T^V, LP SR	10%	Normalisation
`R_HPLP_SR_J0_Whf`	HP to LP ratio	*Whf*, applied to 0-lepton and 1-lepton in HP region	18%	Normalisation
`R_SRCR_CR_Whf`	SR to CR ratio	*Whf*, applied correlated to 0-lepton and 1-lepton, CR	90% (0L) 40% (1L)	Normalisation
`WmJShape`	μ_R var	W+jets, all regions	–	Shape

To study modelling systematic uncertainties for the diboson process, alternative samples are used. The main sources of systematics are the uncertainties on the parameters of the ME calculations and of the PS description. These uncertainties are treated by:

- variation of the renomalisation scale μ_R by a factor 2 or 0.5 with respect to the original value μ_R^{norm} in the ME calculation. The variations are implemented as internal weight in the nominal sample.

Table 6.9 Systematic uncertainties estimated for the Z+jets simulation. The first column states the name of the uncertainty used in the fit, followed by a brief description in the second column. The third column lists the samples the uncertainty is applied to and in which analysis regions, the fourth the value of the uncertainty and the last column states the effect of the uncertainty

Name	Description	Samples/categories	Value	Effect
`norm_Zbb`	*Zhf* norm.	*Zhf*, applied to all the channels, all regions	floating	Normalisation
`ZbbNorm_L0`	*Zhf* norm. ratio	*Zhf*, 0-lepton to 2-lepton ratio, applied to 0-lepton, all regions	16%	Normalisation
`ZclNorm`	*Zcl* norm.	*Zcl*, all regions	30%	Normalisation
`ZlNorm`	*Zl* norm.	*Zl*, all regions	30%	Normalisation
`ZbcZbbRatio`	*Zbc/Zbb* ratio	*Zbc*, applied to 0-lepton and 2-lepton	30%	Normalisation
`ZblZbbRatio`	*Zbl/Zbb* ratio	*Zbl*, applied to 0-lepton and 2-lepton	30%	Normalisation
`ZccZbbRatio`	*Zcc/Zbb* ratio	*Zcc*, applied to 0-lepton and 2-lepton	20%	Normalisation
`R_Ptv_400_Zhf`	med. to high p_T^V ratio	*Zhf*, applied to 0-lepton and 2-lepton, high p_T^V	10%	Normalisation
`R_HPLP_SR_J0_L0_Zhf`	HP to LP ratio	*Zhf*, applied to 0-lepton in HP region	18%	Normalisation
`R_SRCR_CR_Zhf`	SR to CR ratio	*Zhf*, applied to 0-lepton, CR	40%	Normalisation
`ZmJShape`	μ_R var	*Z*+jets, all regions	–	Shape

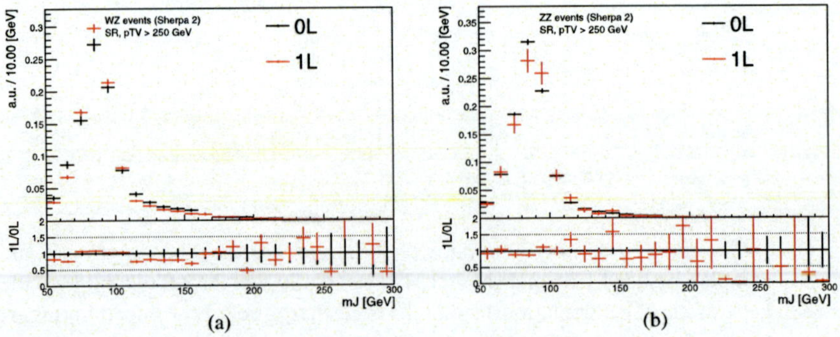

Fig. 6.10 Normalised large-R jet mass distribution of WZ (**a**) and ZZ (**b**) events in the SR with $p_T^V > 250$ GeV in 0- and 1-lepton channels

- variation of the factorization scale μ_F by a factor 2 or 0.5 with respect to the original value μ_F^{norm} in the ME calculation. The variations are implemented as internal weight in the nominal sample.
- alternative sample generated using Powheg for the ME calculation and Pythia for the PS description instead of Sherpa generator. As in the $V+$jets case, large differences between the nominal and the alternative samples are expected from the comparison of the nominal and alternative samples because several parameters of the generator are changed.

For each diboson process a normalisation uncertainty is applied. A normalisation uncertainty, called VZNorm, of 16 and 10% is considered for WZ and ZZ events, while for WW events a normalisation uncertainty, called WWNorm, of 25% is considered.

For the WZ process, in general, the acceptance uncertainties are derived in 1-lepton channel and then applied to all the lepton channels. On the contrary, different approaches are used for the ZZ process. The acceptance uncertainties propagated in the analysis for the VZ process are:

- HP-to-LP relative acceptance uncertainty, called R_HPLP_VZ_SR_J0, of 18% for the WZ and ZZ events applied to the HP SR.
- medium-to-high p_T^V relative acceptance uncertainty, called R_Ptv_VZ_BMin400, of 10% for WZ events. For the ZZ events the medium-to-high acceptance uncertainties of 6 and 18% are considered in the 0-lepton channel and in the other two lepton channels, respectively. The uncertainty is applied in the high p_T^V region.
- channel extrapolation uncertainty of 9% for the WZ events in 0-lepton channel, called WZNorm_L0, of 30% and 18% for ZZ events in 0-lepton and 1-lepton channels respectively. The extrapolation uncertainties for ZZ events are called ZZNorm_L0 and ZZNorm_L1 in 0- and 1-lepton channels, respectively.

For the diboson process no SR-to-CR acceptance uncertainties are considered because the number of diboson events in the CR is low.

The shape uncertainties are derived by adding the contribution of the VZ processes in all the analysis regions. Two shape uncertainties are applied in the analysis:

- a shape uncertainty on m_J, called VVMURMJBoosted, extracted from the comparison of the nominal sample and the alternative sample with varied μ_R. The comparison is done by considering the contribution of all the analysis regions of the three lepton channels. The ratio of the nominal and the alternative samples is fitted with a hyperbolic tangent function (red line in the bottom panel of Fig. 6.11) and then propagated in the final fit.
- a shape uncertainty on m_J, called VVPP8MJBoosted, extracted from the comparison of the nominal Sherpa 2 and alternative Powheg + Pythia 8 samples. The contribution of the 0- and 1-lepton channels are merged. The contribution in the 2-lepton channel is not considered because no significant shape differences are observed between the nominal and the alternative samples due to large statistical fluctuations. The comparison between the two samples in the 0- and 1-lepton

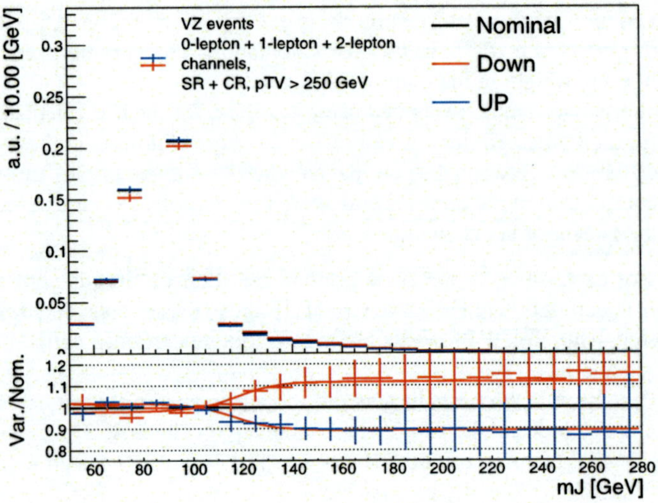

Fig. 6.11 Normalised large-R jet mass distribution of the nominal sample and alternative sample with varied μ_R for WZ and ZZ processes, including the contributions of all the analysis regions. The renormalisation scale μ_R has been varied giving an up and down variation. The *red line* in the *bottom panel* shows the fitting results of the hyperbolic tangent function

channels shows an opposite effect in the LP and HP SR so each region is fitted separately with a third order polynomial function (Fig. 6.12). A protection is applied to the high mass region where a constant function replaces the polynomial function. The blue line in Fig. 6.12 shows the shape uncertainty implemented in the final fit for the HP and LP SR.

Table 6.10 shows a summary of all the modelling systematic uncertainties of the diboson background.

6.3.7 *QCD Multi-jet Modelling*

The QCD multi-jet background represents a subleading reducible background. Due to the large rejection factor of the analysis selection, a description based on MC sample of this background is impossible due to the limited number of events passing the event selection. Data-driven techniques are used to estimate the contribution of mutijet events in the 0- and 1-lepton channels. In the next sub-sections a brief description of the procedures used to extract the QCD multi-jet contribution in 0- and 1-lepton channels.

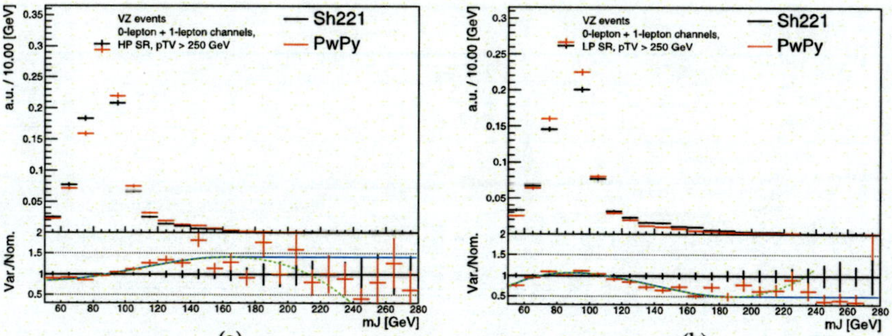

(a) (b)

Fig. 6.12 Normalised large-R jet mass distribution of the nominal Sherpa (*black dots*) and alternative Powheg+Pythia (*red dots*) samples for WZ and ZZ processes in the HP (**a**) and LP (**b**) SRs and summing the contribution of all the other analysis regions of 0- and 1-lepton channels. The renormalisation scale μ_R has been varied giving an up and down variation. The *green dashed line* in the bottom panel shows the fitted third order polynomial functions while the *blue line* shows the function after applying the high mass region protection

Table 6.10 Systematic uncertainties estimated for the diboson simulation. The first column states the name of the uncertainty used in the fit, followed by a brief description in the second column. The third column lists the analysis region in which the uncertainty is applied, the fourth the value of the uncertainty and the last column states the effect of the uncertainty

Name	Description	Samples/Categories	Value	Effect
WWNorm	WW norm.	WW, all channels, all regions	25%	Normalisation
VZNorm	VZ norm.	WZ and ZZ, all channels, all regions	10% for ZZ, 16% for WZ	Normalisation
ZZNorm_L0	ZZ 0L/2L accept.	ZZ, all regions, applied in 0L	30%	Normalisation
ZZNorm_L1	ZZ 1L/2L accept.	ZZ, all regions, applied in 1L	18%	Normalisation
WZNorm_L0	WZ 0L/1L accept.	WZ, all regions, applied in 0L	9%	Normalisation
R_Ptv_VZ_BMin400	med. to high p_T^V ratio	WZ and ZZ, high p_T^V	10% (WZ 0L and 1L), 6% (ZZ 0L), 18% (ZZ 1L and 2L)	Normalisation
R_HPLP_VZ_SR_J0	HP to LP ratio	WZ and ZZ, applied to 0L and 1L in HP signal region	18%	Normalisation
VVMURMJBoosted	μ_R m_J var.	WZ and ZZ, all channels, all regions	-	Shape
VVPP8MJBoosted	PwPy8/Sherpa m_J var.	WZ and ZZ, all channels, all regions	-	Shape

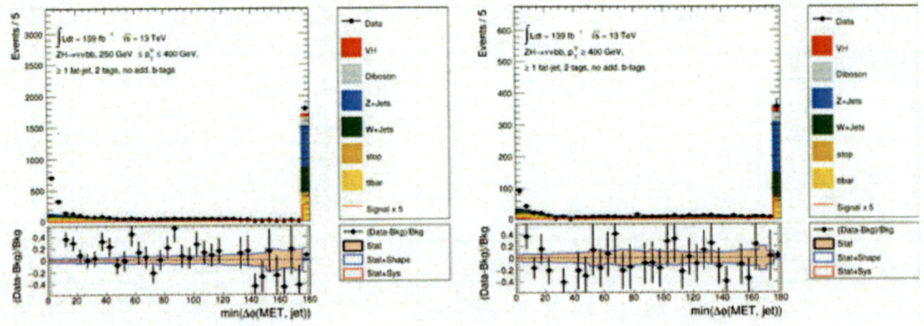

(a) (b)

Fig. 6.13 Pre-fit min[$\Delta\phi(E_{\mathrm{T}}^{\mathrm{miss}}$, small-$R$ jets)] distributions in 250 GeV $\leq p_{\mathrm{T}}^{V} < 400$ GeV (**a**) and $p_{\mathrm{T}}^{V} \geq 400$ GeV (**b**) regions in the SR excluding the selection cut on this variable

QCD Multi-jet in 0-Lepton Channel

Angular cuts between the $E_{\mathrm{T}}^{\mathrm{miss}}$, $E_{\mathrm{T,trk}}^{\mathrm{miss}}$, $H_{\mathrm{cand.}}$ and selected small-R jets (see Sect. 5.2.2) are applied in the 0-lepton channels to suppress the multi-jet and non-collisional background. The data-driven technique used for the MJ estimation is described in the following.

To estimate the MJ contribution, the min[$\Delta\phi(E_{\mathrm{T}}^{\mathrm{miss}}$, small-$R$ jets)] distribution is studied before applying any cut on it. Figure 6.13 shows the min[$\Delta\phi(E_{\mathrm{T}}^{\mathrm{miss}}$, small-$R$ jets)] distribution in the SR in the two p_{T}^{V} bins. In the last bin of the distribution there are all the events with $175° \leq$ min[$\Delta\phi(E_{\mathrm{T}}^{\mathrm{miss}}$, small-$R$ jets)] $< 180°$ and all the events with no small-R jets, with $p_{\mathrm{T}} > 70$ GeV, outside the Higgs candidate jet.[6]

At low min[$\Delta\phi(E_{\mathrm{T}}^{\mathrm{miss}}$, small-$R$ jets)] values, the disagreement between data and the simulated events is assumed to be caused by the MJ background not simulated in the MC samples. To extract the MJ yields, a fit is performed using an exponential decay as functional form to model the MJ background. The exponential model has been tested considering MC di-jets events passing coarser selection criteria than the ones used in the event selection of the analysis.[7] In the fit used to extract the MJ

---

[6] The decision of this high jet p_{T} threshold has been taken in order to not discard signal events with low min[$\Delta\phi(E_{\mathrm{T}}^{\mathrm{miss}}$, small-$R$ jets)] values. Almost 20–30% of signal events in the 0-lepton channel are from the $qq \to WH$ process where the W boson decays in $\tau + \nu$ and the τ lepton decays hadronically. In the $qq \to WH$ process, the jet selected in the min[$\Delta\phi(E_{\mathrm{T}}^{\mathrm{miss}}$, small-$R$ jets)] calculation is the one from the τ lepton decay, it is very soft and it has almost the same direction of the $E_{\mathrm{T}}^{\mathrm{miss}}$ vector. Consequently, applying a low jet p_{T} threshold, i.e. $p_{\mathrm{T}} > 20$ GeV, most of the $qq \to WH$ events in the 0-lepton channel have low angle values. Rising the jet p_{T} cut, the small-R jet arising from the τ lepton decay is not selected in the min[$\Delta\phi(E_{\mathrm{T}}^{\mathrm{miss}}$, small-$R$ jets)] calculation and 10% of the signal events are not reconstructed in the low min[$\Delta\phi(E_{\mathrm{T}}^{\mathrm{miss}}$, small-$R$ jets)] value region.

[7] In this loose event selection, there are no requirements on the number of b-tagged track-jets matched to the Higgs candidate. This means that the 0-tag, 1-tag and 2-tag categories are all considered together.

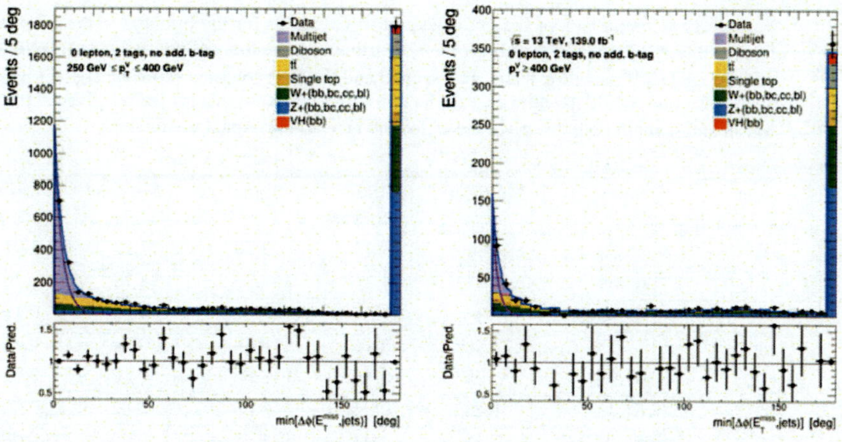

(a) (b)

Fig. 6.14 Post-fit min[$\Delta\phi(E_{\mathrm{T}}^{\mathrm{miss}}$, small-$R$ jets)] distributions in 250 GeV $\leq p_{\mathrm{T}}^{V} < 400$ GeV (**a**) and $p_{\mathrm{T}}^{V} \geq 400$ GeV (**b**) regions in the SR excluding the selection cut on this variable. The MJ is modeled using an exponential shape

contribution, the normalisations of the major backgrounds (the Z+jets, W+jets and $t\bar{t}$) are left floating while the normalisation of the minor backgrounds (diboson and the single top) are fixed to the MC prediction. Figure 6.14 shows the post-fit distributions of min[$\Delta\phi(E_{\mathrm{T}}^{\mathrm{miss}}$, small-$R$ jets)] in the SR in the two p_{T}^{V} bins.

From the fit, the MJ yields and distribution are extracted. Table 6.11 shows the estimate of MJ yields, together with the signal yields, as a function of the cut applied to min[$\Delta\phi(E_{\mathrm{T}}^{\mathrm{miss}}$, small-$R$ jets)] in the SR of the two p_{T}^{V} regions. A cut at 30° on min[$\Delta\phi(E_{\mathrm{T}}^{\mathrm{miss}}$, small-$R$ jets)] is chosen to have only 1% of remaining MJ contamination with respect to the signal. The same cut is applied for both p_{T}^{V} regions. The remaining MJ contribution can be absorbed in the final global fit because the m_J shape of the MJ events is similar to the shape of the other background events.

QCD Multi-jet in 1-Lepton Channel

Also in 1-lepton channel the MJ background is studied using a data-driven method. To evaluate the MJ contribution, a multi-jet enriched control region is defined using inverted lepton isolation cuts. The assumption in the procedure is that the shape of the MJ in the SR is the same as the one in the MJ enriched control region.

The variable chosen to extract the MJ shape is the transverse mass of the W boson m_T^W. To reduce the statistical uncertainties all the studies are performed in the SR without any split in p_{T}^{V} ($p_{\mathrm{T}}^{V} \geq 250$ GeV). Separate MJ templates are extracted depending on the lepton flavour (electron or muon). In the muon sub-channel the statistics in the MJ control region is very low so events in the 0-tag region are

Table 6.11 Comparison of the multi-jet and signal yields for the 0-lepton in the SR determined from a fit to min[$\Delta\phi(E_{\mathrm{T}}^{\mathrm{miss}}$, small-$R$ jets)] distribution. The first column represents the cut applied on min[$\Delta\phi(E_{\mathrm{T}}^{\mathrm{miss}}$, small-$R$ jets)]. The second and the fifth columns show the signal yields for the given selection, while the third and sixth columns show the MJ yields for the given selection. The fourth and seventh columns gives the ratio between MJ and signal yields

	250 GeV < p_{T}^{V} < 400 GeV			p_{T}^{V} > 400 GeV		
Cut	Signal	Multi-jet	Multi-jet/signal	Signal	Multi-jet	Multi-jet/signal
0°	65.80	879.66	13.47	19.74	77.20	3.91
5°	64.69	244.47	4.21	19.36	14.73	0.76
10°	63.66	67.94	1.31	19.02	2.81	0.15
15°	62.73	18.88	0.41	18.88	0.54	0.03
20°	61.98	5.25	0.13	18.66	0.10	0.01
25°	61.43	1.46	0.04	18.56	0.02	0.00
30°	61.02	0.41	0.01	18.48	0.00	0.00
35°	60.65	0.11	0.00	18.40	0.00	0.00
40°	60.30	0.07	0.00	18.32	0.00	0.00

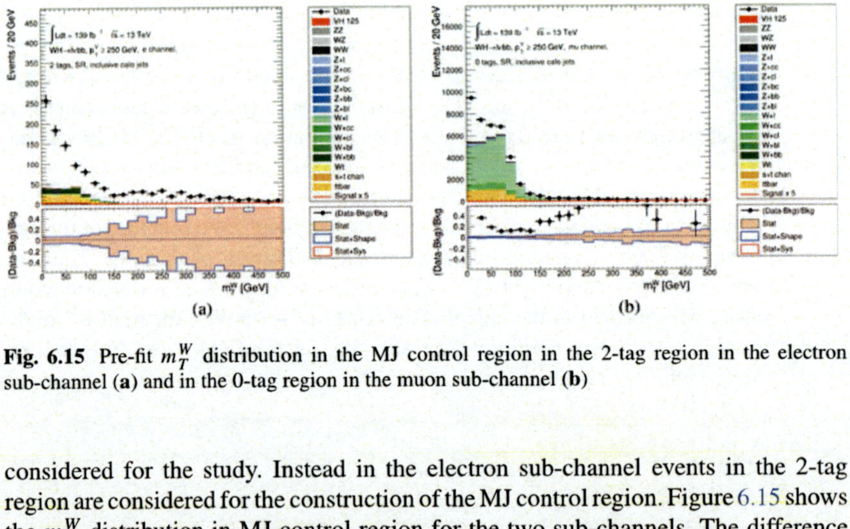

(a) (b)

Fig. 6.15 Pre-fit m_T^W distribution in the MJ control region in the 2-tag region in the electron sub-channel (**a**) and in the 0-tag region in the muon sub-channel (**b**)

considered for the study. Instead in the electron sub-channel events in the 2-tag region are considered for the construction of the MJ control region. Figure 6.15 shows the m_T^W distribution in MJ control region for the two sub-channels. The difference between data and simulated sample is due to the multi-jet contribution.

The shape of the MJ events in the enriched control region is obtained by subtracting the electroweak and top backgrounds ($t\bar{t}$, V + jets, diboson, single top) to the data. A template fit is performed in the SR to extract the MJ yields considering the MJ shape extrapolated in the MJ control region. In the fit the normalisation factors of the top and W+jets backgrounds are extracted simultaneously with the MJ yields. In the electron sub-channel, the MJ fraction in the SR is 2.3% with respect to the total background after applying a cut at $E_{\mathrm{T}}^{\mathrm{miss}}$ > 50 GeV, while in the muon sub-

Table 6.12 Systematic uncertainties estimated for the MJ estimate. The first column states the name of the uncertainty used in the fit, followed by a brief description in the second column. The third column lists the analysis region in which the uncertainty is applied, the fourth the value of the uncertainty and the last column states the effect of the uncertainty. The MJ systematic uncertainty is uncorrelated in all the analysis regions of the 1-lepton channel

Name	Description	Samples/categories	Value	Effect
`QCDMJE1_L1`	MJ norm.	MJ electron, 1-lepton, decorrelated in all regions	55%	Normalisation

channels the fraction of MJ is less than 0.5% without any cut on $E_{\mathrm{T}}^{\mathrm{miss}}$. For this reason no uncertainties are considered in the muon sub-channel, while in the electron sub-channel, a normalisation uncertainty of 55%, called `QCDMJE1_L1`, has been applied in the fit. The MJ systematic uncertainty is treated uncorrelated in all the analysis regions of the 1-lepton channel. Table 6.12 shows the systematic model for the MJ background.

References

1. ATLAS Collaboration (2019) Luminosity determination in pp collisions at $\sqrt{s} = 13 TeV$ using the ATLAS detector at the LHC, ATLAS-CONF-2019-021. https://cds.cern.ch/record/2677054
2. ATLAS Collaboration (2016) Performance of b-jet identification in the ATLAS experiment, JINST 11: P04008. https://doi.org/10.1088/1748-0221/11/04/P04008, arXiv: 1512.01094 [hep-ex]
3. ATLAS Collaboration (2018) Evaluation of theoretical uncertainties for simplified template cross section measurements of V-associated production of the Higgs boson, ATL-PHYS-PUB-2018-035. https://cds.cern.ch/record/2649241
4. Frixione S, Laenen E, Motylinski P, Webber BR, White CD (2008) Single-top hadroproduction in association with a W boson. JHEP 07:029. https://doi.org/10.1088/1126-6708/2008/07/029 arXiv: 0805.3067 [hep-ph]

Chapter 7
Statistical Treatment

This chapter contains a detailed description of the statistical procedures implemented in the analysis to extract the final results. The likelihood function and the statistical procedures used to test the fit results are described. In addition this chapter illustrates the tools used to scrutinize and validate the fit results.

7.1 General Statistical Treatment

7.1.1 The Likelihood Function

To measure the signal yield, a binned maximum likelihood fit is performed. The observable used in the fit as final discriminant is the large-R jet mass m_J. The fit is done simultaneously to all the analysis regions and the three lepton channels extracting the VH and VZ contributions. The likelihood is defined as the product over all bins of the Poisson probability to observe n_i events when m_i events are expected in a certain bin i:

$$\begin{aligned}\mathcal{L}_{Pois}(\boldsymbol{\mu}, \boldsymbol{\alpha}, \boldsymbol{\gamma}, \boldsymbol{\tau}) &= \prod_{i \in \text{bins}} \text{Pois}(n_i | m_i(\boldsymbol{\mu}, \boldsymbol{\alpha}, \boldsymbol{\gamma}, \boldsymbol{\tau})) \\ &= \prod_{i \in \text{bins}} \text{Pois}(n_i | \mu_{VH} s_i(\boldsymbol{\alpha}) + \mu_{VZ} b_i^{VZ}(\boldsymbol{\alpha}, \boldsymbol{\gamma}, \boldsymbol{\tau}) + b_i^{oth}(\boldsymbol{\alpha}, \boldsymbol{\gamma}, \boldsymbol{\tau})) \\ &= \prod_{i \in \text{bins}} \frac{(\mu_{VH} s_i + \mu_{VZ} b_i^{VZ} + b_i^{oth})^{n_i}}{n_i!} e^{-(\mu_{VH} s_i + \mu_{VZ} b_i^{VZ} + b_i^{oth})}\end{aligned} \tag{7.1}$$

where $\boldsymbol{\mu} = (\mu_{VH}, \mu_{VZ})$ and the number of expected events m_i in bin i is obtained summing the expected signal s_i and background events b_i. The contribution of the

G. Di Gregorio, *A Beauty-ful Boson*, Springer Theses,
https://doi.org/10.1007/978-3-031-20013-7_7

expected background is split in contribution of the VZ background b_i^{VZ} and contribution of the remaining backgrounds b_i^{oth} ($b_i = \mu_{VZ} b_i^{VZ} + b_i^{oth}$). The expected signal s_i and VZ background b_i^{VZ} events are multiplied by the parameters μ_{VH} and μ_{VZ}, respectively. The parameter μ_{VH} (μ_{VZ}) is referred to as *signal strength* and it is defined as the ratio of the measured cross-section time branching ratios $\sigma \times$BR for the VH (VZ) process divided by its SM expectation. The signal strength parameters are also called Parameter of Interests (PoIs). In the analysis presented in this thesis, a multi-PoIs fit is performed since the values of μ_{VH} and μ_{VZ} signal strengths are extracted simultaneously. In particular this convention is used: the (x+y) PoIs fit indicates a simultaneous fit in which x is the number of VH PoIs and y the number of VZ PoIs. This means that the multi-PoIs fit described in the following is a (1+1) PoIs fit.

In addition to the parameters μ_{VH} and μ_{VZ}, the likelihood depends on other parameters $\boldsymbol{\alpha}, \boldsymbol{\gamma}, \boldsymbol{\tau}$ called Nuisance Parameters (NPs). The NPs encode the dependence of the prediction on the systematic uncertainties into continuous parameter in the likelihood. The NPs can be categorized into three classes: $\boldsymbol{\alpha} = (\alpha_1, \alpha_2, ..)$, $\boldsymbol{\gamma} = (\gamma_1, \gamma_2, ..)$ and $\boldsymbol{\tau} = (\tau_1, \tau_2, ..)$. The $\boldsymbol{\tau}$ NPs are unconstrained parameters controlling the normalisation of the backgrounds and they are called *free-floating* because they are free to float in the fit. The $\boldsymbol{\gamma}$ NPs represent the statistical uncertainties caused by the limited size of simulated background samples. The signal process is usually chosen to not be affected by the $\boldsymbol{\gamma}$ NPs as the statistical uncertainties on the predicted signal simulation are small with respect to the backgrounds. A γ NP is applied in each bin of the analysis on the sum of all the backgrounds. The modelling and the experimental uncertainties enter in the fit through the $\boldsymbol{\alpha}$ NPs and they affect both signal and background events.

The $\boldsymbol{\alpha}$ NPs are estimated from data or auxiliary measurements which provide both central values and uncertainties. For each $\boldsymbol{\alpha}$ NP, the likelihood function is multiplied by an *auxiliary term* that constrains the value of the systematic uncertainty around its estimate, within the uncertainty on such estimate. The auxiliary terms are Gaussian functions with mean equal to zero and variance equal to one:

$$\mathcal{L}_{aux}(\boldsymbol{\alpha}) = \prod_{\alpha \in \boldsymbol{\alpha}} \text{Gauss}(\boldsymbol{\alpha}|0, 1) = \prod_{\alpha \in \boldsymbol{\alpha}} \frac{1}{\sqrt{2\pi}} e^{-\alpha^2/2} \tag{7.2}$$

where the product is extended over all the systematic uncertainties considered in the analysis. The NPs are defined such that for $\alpha_j = 0$ the nominal predictions are obtained and for $\alpha_j = \pm 1$ two modified templates, called *up/down template*, for the $\pm 1\sigma$ variation are obtained. The NPs are expressed in unit of their uncertainties σ_α.[1]

The uncertainties on the background predictions due to the limited number of simulated events are also accounted in the likelihood function considering Poisson terms:

[1] This means, for example, that moving the parameter α of the jet energy scale by 1 corresponds in shifting the energy calibration of the jet by 1σ.

$$\mathcal{L}_{Stat}(\boldsymbol{\gamma}) = \prod_{i \in \text{bins}} \frac{(\gamma_i b_i)^{b_i} e^{-(\gamma_i b_i)}}{\Gamma(b_i + i)} \tag{7.3}$$

where

$$\Gamma(x) = \int dt\, t^{x-1} e^{-t} \tag{7.4}$$

For each bin i of the histograms a γ_i NP is considered and it represents the uncertainty on the sum of all the background processes in that bin. Dedicated studies have shown that, while the statistical uncertainty of the VH sample is small, the one of VZ is much more in-line with the other backgrounds. Consequently, in the multi-PoIs fit the VZ is treated like all the other backgrounds for the γ parameters.

The full likelihood used in the final fit can be schematically written as:

$$\mathcal{L}(\mu, \boldsymbol{\alpha}, \boldsymbol{\gamma}, \boldsymbol{\tau}) = \mathcal{L}_{Pois}(\mu, \boldsymbol{\alpha}, \boldsymbol{\gamma}, \boldsymbol{\tau}) \cdot \mathcal{L}_{aux}(\boldsymbol{\alpha}) \cdot \mathcal{L}_{Stat}(\boldsymbol{\gamma}) \tag{7.5}$$

A binned likelihood fit is performed to determine the PoIs and their uncertainties. The measured signal strengths and the NPs are obtained as the values of the parameters that maximize the likelihood function $\mathcal{L}(\mu, \boldsymbol{\alpha}, \boldsymbol{\gamma}, \boldsymbol{\tau}) = \mathcal{L}(\mu, \boldsymbol{\theta})$ or, equivalently, minimize $-\ln \mathcal{L}(\mu, \boldsymbol{\theta})$, where $\boldsymbol{\theta}$ represents the set of NPs introduced previously, $\boldsymbol{\theta} = (\boldsymbol{\alpha}, \boldsymbol{\gamma}, \boldsymbol{\tau})$. The likelihood maximization without fixing the values of signal strengths is called *unconditional fit*. Instead, the *conditional fit* is performed maximising the likelihood for particular values of the signal strengths.

The uncertainties on the signal strengths is determined evaluating $\mu_{+/-} = \hat{\mu}^{+\sigma_\mu^+}_{-\sigma_\mu^-}$ as the value that satisfies this equation:

$$-2 \ln \frac{\mathcal{L}(\mu_{+/-}, \hat{\hat{\boldsymbol{\theta}}})}{\mathcal{L}(\hat{\mu}, \hat{\boldsymbol{\theta}})} = 1 \tag{7.6}$$

where $\hat{\mu}$ and $\hat{\boldsymbol{\theta}}$ are the parameters that maximise the overall likelihood and $\hat{\hat{\boldsymbol{\theta}}}$ are the NP values that maximise the likelihood for a particular value of μ.

7.1.2 *Profile Likelihood Ratio and Test Statistic*

The Profile Likelihood Ratio (PLR) $\lambda(\mu)$ is defined as the ratio of two Likelihood functions:

$$\lambda(\mu) = \frac{\mathcal{L}(\mu, \hat{\hat{\boldsymbol{\theta}}}(\mu))}{\mathcal{L}(\hat{\mu}, \hat{\boldsymbol{\theta}})} \tag{7.7}$$

where $\hat{\mu}$ and $\hat{\boldsymbol{\theta}}$ are the parameters that maximise the overall likelihood and $\hat{\hat{\boldsymbol{\theta}}}(\mu)$ are the NP values that maximise the likelihood for a particular value of μ as mentioned previously. The PLR is used to construct the *test statistic* $q_\mu = -2\ln\lambda(\mu)$, which differentiates the background-only hypothesis with $\mu = 0$ from the alternative hypothesis with $\mu > 0$. The PLR takes values $0 \le \lambda(\mu) \le 1$, where large values close to unity imply good agreement between the hypothesised signal strengths μ and the observed data.

A test statistic used in the $VH(b\bar{b})$ analysis is the one for the discovery of a positive signal in which the background-only hypothesis with $\mu = 0$ is tested. The compatibility of the data with the background-only hypothesis is evaluated from the test statistic $q_0 = -2\ln\lambda(\mu = 0) = -2\ln\frac{\mathcal{L}(\mu=0,\hat{\hat{\theta}}(\mu=0))}{\mathcal{L}(\hat{\mu},\hat{\theta})}$. If the data are compatible with the background-only hypothesis, the nominator and the denominator of the test statistic are similar and q_0 is close to 0. Differently, if the event yield is larger than the expectation, the test statistics q_0 assumes larger values indicating higher incompatibility between the data and the tested hypothesis. The incompatibility can be expressed with the p-value, in this case named p_0, defined as:

$$p_0 = \int_{q_{o,obs}}^{\infty} f(q_0|0)dq_0 \tag{7.8}$$

where $q_{o,obs}$ is the value of the test statistic measured from the observed data and $f(q_0|0)$ is the probability density function of the test statistic q_0 under the $\mu = 0$ assumption. The hypothesis of a test can be considered excluded if its p-value is observed below a specific threshold. The p-value can be expressed in terms of the significance Z which is defined such that a Gaussian distributed variable found Z standard deviations above its mean has an upper-tail probability equal to p_0 (Fig. 7.1). In a more formal way, the significance is the quantile (inverse of the cumulative distribution) of the standard Gaussian, computed for $(1 - p_0)$:

$$Z = \Phi^{-1}(1 - p_0) \tag{7.9}$$

The rejection of the background hypothesis with a significance of at least Z = 5 (which correspond to $p_0 = 2.87 \times 10^{-7}$) is considered as an appropriate level to quote a discovery.

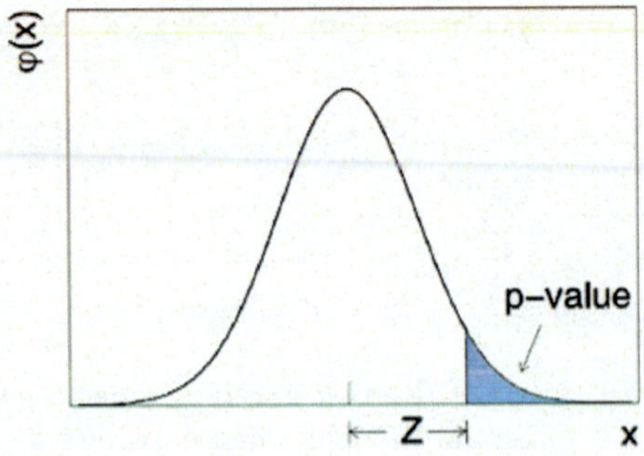

Fig. 7.1 Relationship between a p-value and a significance of Z sigma [3]

To compute the p-value, the distribution of $f(q_0|0)$ is required. This can be achieved by sampling the distribution exploiting the Monte Carlo method [1]. However, the procedure is computationally expensive and approximate solutions are adopted. Assuming the null hypothesis to be true, the Wilk's theorem [2] ensures that q_0 is asymptotically distributed as a χ^2 with one degree of freedom. This means that the value of q_0 can be easily compared to the χ^2 value. With few steps it can be shown that the significance can be $Z = \sqrt{q_0}$. In the following, all the statistical tests are done using the asymptotic approximation ensured by the Wilk's theorem.

7.2 Fit Input

The signal and control regions used in the fit have been summarised in Table 5.5. The following processes are considered in the fit, either as signal or backgrounds:

- signal VH, $H \to b\bar{b}$ (summed over all the production modes);
- Z+jets and W+jets. The V+jets backgrounds are split into three different components depending the flavour composition of the two jets used to reconstruct the Higgs boson decay, V+HF, $V + cl$, $V + ll$;
- $t\bar{t}$;
- single-top: s-, t- and Wt-channels. The s- and t-channels are treated as one component, while the Wt-channel is treated independently;
- diboson: WW, ZZ, WZ. The WZ and ZZ processes are treated as one component and they have an associated PoI in the fit;
- multi-jet in 1-lepton channel. The multi-jet contribution in 0- and 2-lepton channel is negligible.

Signal and background m_J templates are determined from the MC simulation in all the cases except fot the multi-jet background in the 1-lepton channel, whose contribution is extracted from the data.

The likelihood is built from the m_J histograms for each process listed above. The choice of using different binnings and ranges for the m_J distribution has been made to maximise the resolution taking into account the following aspects:

- avoid empty bins in the templates;
- minimise empty bins in the data distributions;
- have a statistical uncertainty in each bins lower than 20% to avoid potential biases on μ.

7.3 Nuisance Parameters

The impact of all the experimental and modelling uncertainties affecting the m_J templates is quantified using histograms that correspond to $\pm 1\sigma$ variation of each specific NP. The up $(+1\sigma)$ and down (-1σ) variations are calculated relative to the

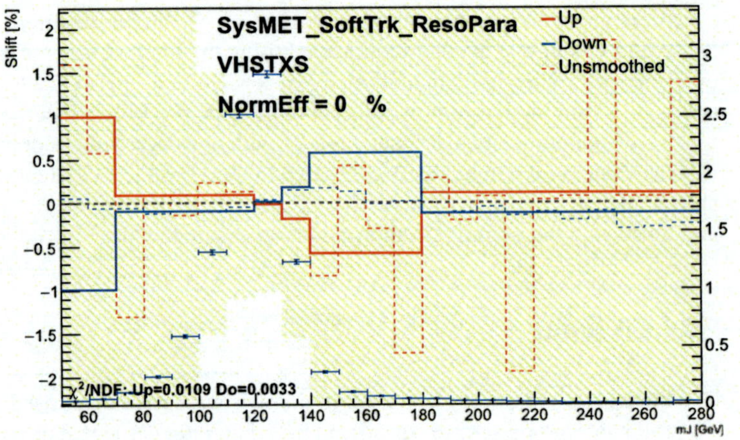

Fig. 7.2 Effect of the smoothing procedure on the m_J distribution for the signal sample in LP SR, $p_T^V \geq 400$ GeV in 1-lepton channel

nominal template. There are few cases where the systematic variation lead only to $+(-)1\sigma$ effect. The jet resolution uncertainty, for example, is a one-sided uncertainty and consequently its effect is symmetrised with respect to the nominal histogram in order to have also the variation in the other direction. In all these cases the effect of the systematic is symmetrized with respect to the nominal histogram to obtained the variation also in the other direction.

Smoothing and Pruning of the Systematic Uncertainties
Certain systematic uncertainties such as the large-R jet energy scale uncertainties can cause bin migrations of events in the m_J distribution. This migration of the events causes large statistical fluctuations which are not physical. To prevent these effects, the *smoothing* procedure is adopted to all the systematic variation across all regions. In the first step, bins are merged until there is one maximum in the varied distribution relative to the nominal distribution. In the second step, bins are furthermore grouped until the statistical uncertainty in each bin is below 5%. Figure 7.2 shows the effect of the smoothing on a E_T^{miss} systematic for the signal VH sample. The red and blue lines show the shifts of the variation of the systematic with respect to the nominal (referred to the left y-label). The dotted lines represents the variations of the systematic before the smoothing while the continuous lines the variations of the systematic after the smoothing procedure. The points with the error bars show the m_J distribution of the signal sample, referring to the right y-label.

From the total list of NPs, only some of them have a sizeable impact on the fit templates. To reduce the number of NPs in order to obtain a more solid fit model, a *pruning* procedure is applied. The procedure removes systematics uncertainties that have a negligible impact on the final result. Normalisation and shape uncertainties are dropped if the variation of the corresponding template is below 0.5% in all the bins.

Additional pruning criteria are applied in all the analysis regions where the signal contribution is less than 2% of the total background and the systematic variation impact of the total background is less than 0.5%.

7.4 Tool for the Validation of the Fit Results

All the tools described in the following are used to understand the statistical analysis and to identify potential problems and errors in the fit.

7.4.1 Pull Plots

The information of the fit results can be visualised using plots. The *pull plots* shows the pull of the nuisance parameters which is the comparison of the central value and uncertainty of the nuisance parameters before and after the fit [3]. The pull of the NP θ, with an expectation value θ_0 and standard deviation σ_θ, is defined as:

$$pull(\theta) = \frac{\hat{\theta} - \theta_0}{\sigma_\theta} \tag{7.10}$$

where $\hat{\theta}$ is the NP value obtained from the maximum likelihood fit. The pull quantifies how far from its expected value the NP is "pulled" by the fit in number of σ_θ. A healthy situation is when the pull is zero, if this is not the case, further investigation is required. If the pull is not zero, the NP value extracted from the fit is different from the expected NP value.

In the pull plots the parameters corresponding to the floating normalisations are also shown but following a different convention. The value shown in the plot is not the pull since the floating normalisations do not have any prior, but it is the absolute value of the normalisation with its uncertainty.

A possible estimate of the error of the NP can be performed studying the PLR as function of the parameter θ around $\hat{\theta}$. The estimate is done applying the same method used to evaluate the μ uncertainty. For the NPs with a Gaussian constraint in the likelihood, the expected interval of the pull is $[-1, +1]$. If the interval is smaller than the expected one, the performed measurement is more accurate with respect to the auxiliary measurement. In this case, the systematic uncertainty is "constrained" by the data and it needs to be understood.

Once all the systematics are considered inside the fit, the first fit is performed using the *Asimov dataset*. The *Asimov dataset* [2] has as data the expected yields predicted from the simulation and they are used in replacement of the real data to test the fit performance and to quantify the expected sensitivity. By definition, the value of the pull from the fit to the Asimov dataset will not change, it can be only constrained.

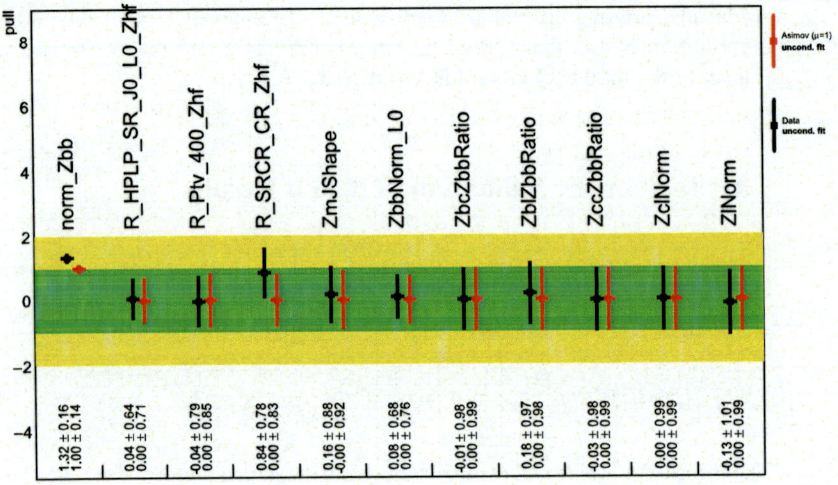

Fig. 7.3 Example of pull plot obtained from an unconditional fit to Asimov (*red dots*) and real dataset (*black dots*)

The pulls of NPs with a Gaussian constraints are set to zero, while the values of the floating normalisations are equal to 1. The use of the Asimov dataset is important to spot suspicious behaviours and to predict the expected precision of the floating normalisation factors. Figure 7.3 shows an example of pull plot in which there is a comparison of the pulls obtained from the unconditional fit applied to Asimov (red dots) and to real dataset (black dots).

Another way to study the stability of the fit without using the information of the PoIs is to perform a conditional fit with $\mu = 1$. With this fit the value of the PoIs is fixed but it is possible to extract information on the pulled NPs.

7.4.2 Correlation Matrix

Another tool used for the validation of the fit model is the correlation matrix. The correlation between θ_i and θ_j NPs or PoIs is obtained from the covariance matrix of the estimator of all the parameters, $V_{ij} = \text{cov}[\hat{\theta}_i, \hat{\theta}_j]$. In the large sample limit, the covariance matrix is defined as the inverse of the second derivative of the log-likelihood function evaluated at $\hat{\mu}$ and $\hat{\theta}$ [2]:

$$\text{cov}[\theta_i, \theta_j] = \left[-\frac{\partial^2 \ln \mathcal{L}(\theta)}{\partial\theta_i \partial\theta_j} \right]^{-1} \tag{7.11}$$

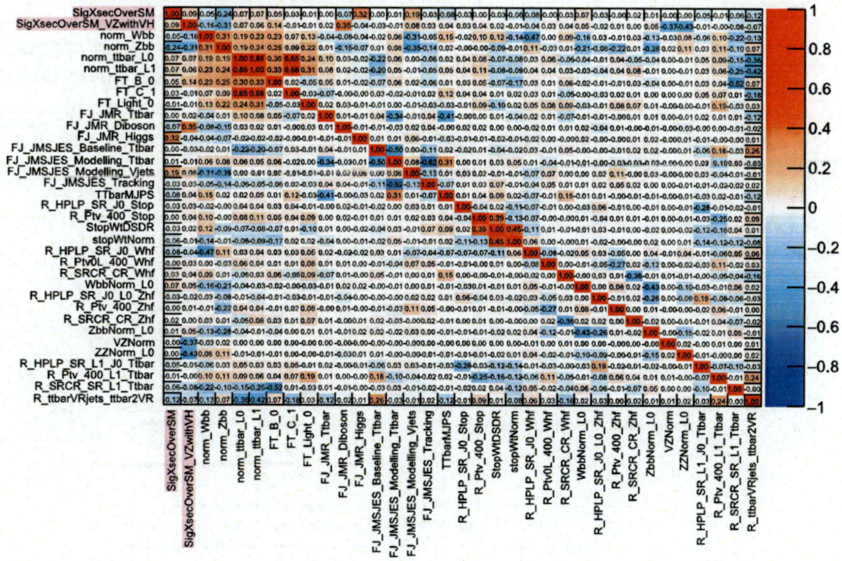

Fig. 7.4 Correlation matrix obtained from a fit to the Asimov dataset

The value of the correlation coefficients can vary from -1 to $+1$. If two variables are not related, their correlation is zero. Figure 7.4 shows an example of correlation matrix obtained from a fit to the Asimov dataset. The correlation matrix contains also the correlation coefficients between the PoIs (highlighted in magenta) and the NPs. Since some NPs are correlated, the correlation matrix helps to understand why some NPs are constrained or pulled. To simplify a bit the plot, only NPs that have the absolute value of the correlation of magnitude 0.25 or higher with an other NP are shown. In general the NPs have a small correlation with the exception of few cases as the correlation among the normalisation factors and the correlation among the large-R jet systematics (see Fig. 7.4).

7.4.3 Ranking Plot

An important information of the fit is how much the PoI value varies when changing the value of an NP. The impact of a NP θ on the fitted PoI is defined as [3]:

$$\text{impact} = \Delta\mu^{\pm} = \hat{\hat{\mu}}_{\theta_0 \pm \sigma_\theta} - \hat{\mu} \tag{7.12}$$

where $\hat{\mu}$ is the value maximising the likelihood and $\hat{\hat{\mu}}_{\theta_0 \pm \sigma_\theta}$ is the value of the PoI extracted from a fit where all the NPs are allowed to vary except for θ which is fixed to the values at the edge of the intervals of the pulls, $\theta_0 \pm \sigma_\theta$. Each NP has its impact and not all the NPs are equally important. The *ranking plot* is used to sort the NPs with the largest impact. Figure 7.5 shows an example of ranking plot obtained from a fit to the Asimov dataset, in which all the uncertainties are listed in a decreasing order of their impact. The plot shows only 15 NPs with the highest

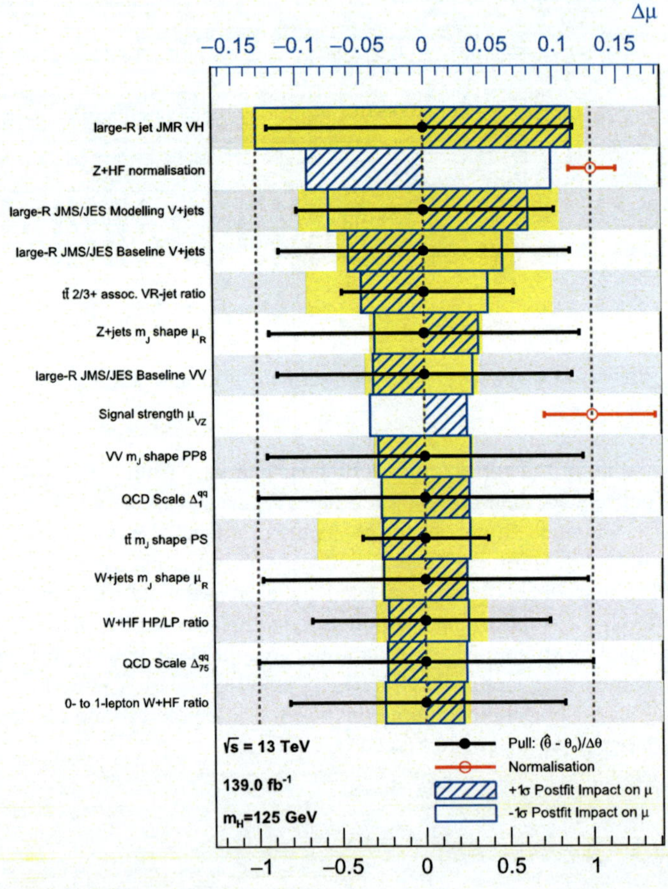

Fig. 7.5 Impact of the uncertainties on the VH signal strength μ_{VH} parameter obtained from a fit to the Asimov dataset. The uncertainties are sorted in a decreasing order. The *boxes* show the variation of $\hat{\mu}_{VH}$, referring to the top x-axis, when fixing the corresponding NP. The impact of the up- and down-variations can be distinguished via the dashed and plane box fillings. The *yellow boxes* show the pre-fit impact, referring to the top x-axis, by varying each NP by $\pm 1\sigma$. The filled black points with the corresponding error bars show the pull of each NP, referring to the bottom x-axis. The *open red circles* with the error bars show the fitted values and the uncertainties of the normalisation factors which are freely floating in the fit. The *dotted vertical* are placed at ± 1 and are referred to the bottom x-axis

impact. The boxes show the impact of the NP on $\hat{\mu}_{VH}$, referring to the top x-axis. The hatched and plane boxes represent the up- and down-variation, respectively. The yellow band shows the pre-fit impact, referring to the top x-axis, by varying each nuisance parameter by $\pm 1\sigma$. The filled black circles with the error bars show the pulls of the NPs and their uncertainties, referring to the bottom x-axis. The open red circles with the error bars show the fitted values and uncertainties of the floating normalisations. By definition of fit to Asimov dataset, all the black points are set to zero and all the red points are set to one. The dotted vertical lines are referred to the bottom x-axis and placed at ± 1. In the ranking plot shown in Fig. 7.5 the highest ranked parameters are the parameters that shows the largest correlation to the signal strength.

7.4.4 Breakdown of the Uncertainties

The uncertainties with similar origin can be grouped together to study the uncertainty impact of the group on the fitted signal strength. In this way it is possible to find which systematics have a big impact on the measurement precision. The uncertainty impact of a group of uncertainties is the result of the comparison of the uncertainties on the signal strengths:

$$\text{uncertainty impact} = \sqrt{\sigma^2_{\hat{\mu}} - \sigma^2_{\hat{\mu}'}} \tag{7.13}$$

where $\sigma^2_{\hat{\mu}}$ is the uncertainty on the signal strength obtained from the nominal fit[2] and $\sigma^2_{\hat{\mu}'}$ is the uncertainty on the signal strength running a fit with all the NPs belonging to a group fixed to their best fit values. When testing the impact of the systematic on one PoI, the understudy PoI is fixed to the value extracted from the nominal fit while the other PoI is left floating in the fit. The "total statical" impact is evaluated comparing the result of the nominal fit with the result of the fit with all the NPs fixed to their best fit values except for the floating normalisations. The "data stat only" impact is defined as the comparison between the nominal fit and the fit with all the NPs fixed to their nominal expectation values. The "floating normalisation" contribution is the quadratic difference between the total error and the error from the fit with only the normalisation factors fixed to the best fit values. The "total systematic" impact is the quadratic difference between the total error and the "total statistical" error. The sum in quadrature of the individual contributions of the systematic uncertainties differs from the total systematic contribution due to correlations between the NPs.

---

[2] The nominal fit indicates the maximum likelihood fit in which all the NPs and PoIs are left floating.

References

1. Metropolis N, Ulam S (1949), The Monte Carlo method. J Am Stat Assoc 44: 335. https://doi.org/10.1080/01621459.1949.10483310
2. Cowan G, Cranmer K, Gross E, Vitells O (2011) Asymptotic formulae for likelihood-based tests of new physics. Eur Phys J C 71:1554. https://doi.org/10.1140/epjc/s10052-011-1554-0 arXiv:1007.1727 [physics.data-an], Erratum: Eur Phys J C 73: 2501 (2013)
3. Gross E (2018) Practical statistics for High Energy Physics. In: CERN Yellow Reports. School Proceedings, vol 3, ed. by Mulders M, Zanderighi G 199, https://doi.org/10.23730/CYRSP-2018-003.199

Chapter 8
Results of the $VH(b\bar{b})$ Boosted Analysis

This chapter shows the results obtained with the $VH(b\bar{b})$ boosted analysis using the full Run 2 dataset considering the large-R *jet mass as final discriminant. The first part describes the results obtained extracting the signal strength as parameter of interest of the fit, while the second part presents the $VH(b\bar{b})$ cross-section measurements together with the constraints on anomalous couplings in the framework of Standard Model effective field theory.*

8.1 Results of the Simultaneous Fit to VH and VZ

8.1.1 Signal Strengths and Significances

A binned maximum-profile likelihood fit is performed to data collected during the Run 2, corresponding to an integrated luminosity of 139 fb^{-1}, in all the ten SRs and four CRs. The final discriminant is the large-R jet mass m_J. From the fit, two signal strengths are extracted simultaneously μ_{VH} and μ_{VZ}. The measured signal strengths for the VH and VZ processes are:

$$\begin{aligned} \mu_{VH} &= 0.72^{+0.39}_{-0.36} = 0.72^{+0.29}_{-0.28}(\text{stat.})^{+0.26}_{-0.22}(\text{syst.}) \\ \mu_{VZ} &= 0.91^{+0.29}_{-0.23} = 0.91 \pm 0.15(\text{stat.})^{+0.25}_{-0.17}(\text{syst.}) \end{aligned} \tag{8.1}$$

The results are well compatible with the predictions of the SM. In the VH result the systematic uncertainties are almost equal to the total statistical error, while in the VZ result the largest uncertainties have a systematic source. A detailed description of the major contributions to the uncertainties on the signal strength is reported in Sect. 8.1.5. The correlation between μ_{VH} and μ_{VZ} is approximately 11%.

G. Di Gregorio, *A Beauty-ful Boson*, Springer Theses,
https://doi.org/10.1007/978-3-031-20013-7_8

Table 8.1 Expected and observed significance values for the VH and VZ processes

	Expected significance (σ)	Observed significance (σ)
VH process	2.7	2.1
VZ process	5.7	5.4

Table 8.2 Normalisation factors for $t\bar{t}$, W+HF and Z+HF backgrounds obtained from the nominal fit. The errors include the statistical and systematic uncertainties

Process	Normalisation factor
$t\bar{t}$ in 0-lepton	0.88 ± 0.10
$t\bar{t}$ in 1-lepton	0.83 ± 0.09
W+HF	1.12 ± 0.14
Z+HF	1.32 ± 0.16

The expected and observed significances obtained from the simultaneous fit are presented in Table 8.1. The expected significance is extracted from a conditional fit to data fixing $\mu_{VZ} = \mu_{VH} = 1$. For a Higgs boson mass of $m_H = 125$ GeV, the observed excess with respect to the background-only hypothesis has a significance of 2.1 standard deviations. Differently, for the VZ process a significance of 5.2 standard deviations is observed providing a direct observation of the $VZ(b\bar{b})$ production mode.

The normalisation factors of the largest backgrounds ($t\bar{t}$, W+HF, Z+HF) are summarized in Table 8.2. Two normalisation factors are extracted for the $t\bar{t}$ background, one in 0-lepton channel and the other in 1-lepton channel, because in the analysis there are two separate top-CRs. The fit tends to reduce the estimated contribution of $t\bar{t}$ events and to increase the contribution of V+HF events. All the normalisation factors agree with the SM predictions within their uncertainties.

8.1.2 *Post-fit m_J Distribution*

The post-fit large-R jet mass m_J distributions with signal strengths, background normalisations and all the NPs set to their best-fit values, are shown in Figs. 8.1, 8.2 and 8.3 for the three lepton channels. A good agreement between data and the predictions is observed in all the SRs and CRs.

To better visualize the Z and Higgs boson peak, all the backgrounds except for the VZ diboson process have been subtracted from the data. Figure 8.4 shows the large-R jet mass m_J distribution summed over all the three lepton channels and SRs. The three channels are weighted by their respective values of the ratio of the fitted Higgs boson signal to backgrounds yields, after subtraction of all the backgrounds

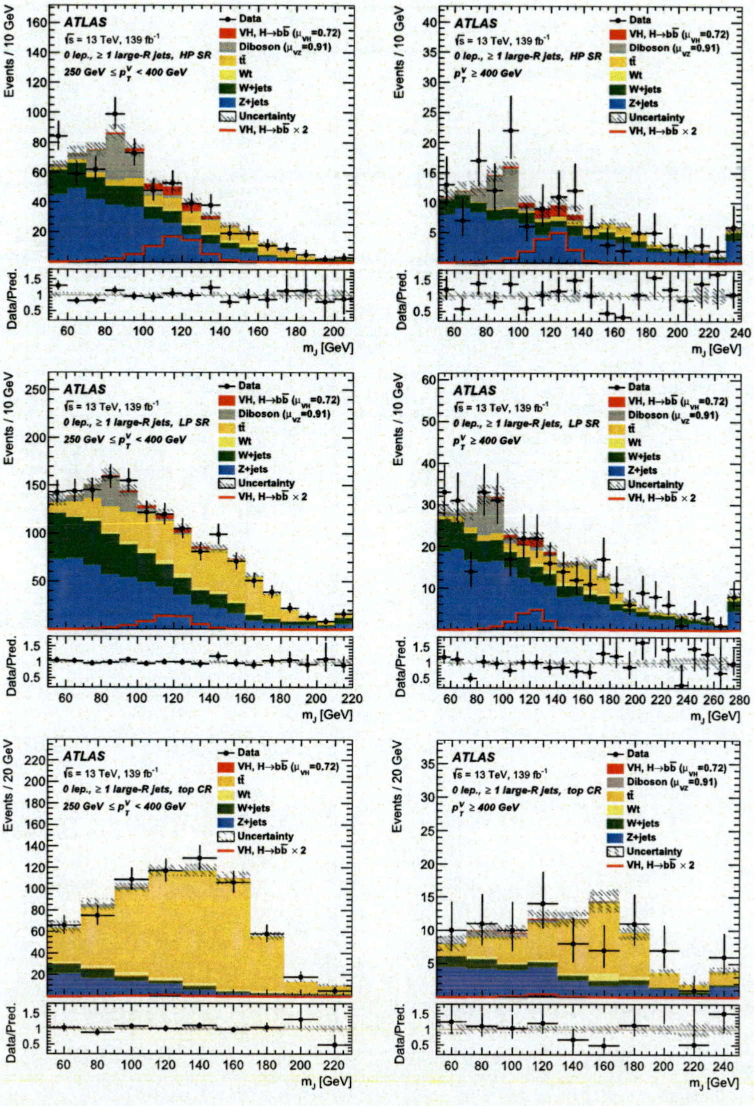

Fig. 8.1 The large-R jet mass post-fit distribution in 1-lepton channel in the HP SRs (*top row*), LP SRs (*middle row*) and CRs (*bottom row*) for 250 GeV$\leq p_{\mathrm{T}}^{V} < 400$ GeV (left) and $p_{\mathrm{T}}^{V} \geq 400$ GeV (right) regions. The data are shown as *black dots*. The background contributions are shown as filled histograms. The Higgs boson signal with $m_H = 125$ GeV is shown as a filled histogram on the top of the fitted background multiplied by the signal strength μ_{VH}, and unstacked as an unfilled histogram, multiplied by a factor 2. The size of the combined statistical and systematic uncertainty for the sum of the signal and background is indicated by the hatched band. The highest bin in the distributions contains the overflow. The ratio of the data to the sum of the fitted signal and background is shown in the *lower panel* [1]

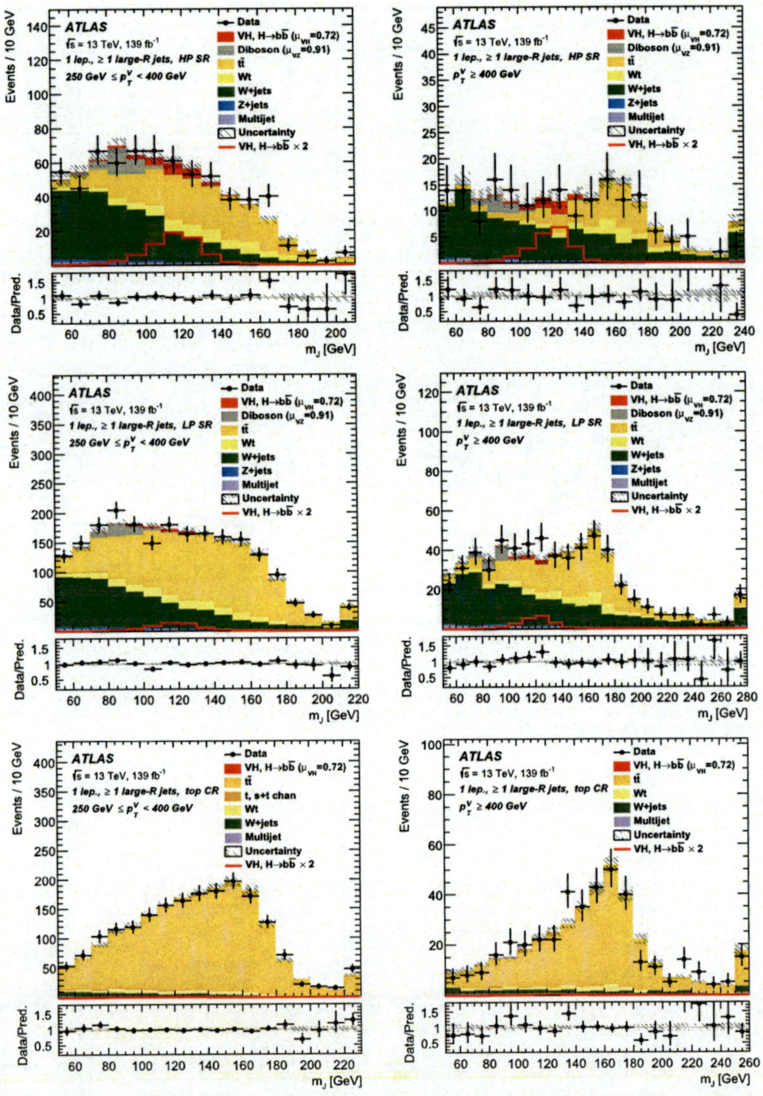

Fig. 8.2 The large-R jet mass post-fit distribution in 1-lepton channel in the HP SRs (*top row*), LP SRs (*middle row*) and CRs (*bottom row*) for 250 GeV$\leq p_{\mathrm{T}}^{V} < 400$ GeV (left) and $p_{\mathrm{T}}^{V} \geq 400$ GeV (right) regions. The data are shown as *black dots*. The background contributions are shown as filled histograms. The Higgs boson signal with $m_H = 125\,\mathrm{GeV}$ is shown as a filled histogram on the top of the fitted background multiplied by the signal strength μ_{VH}, and unstacked as an unfilled histogram, multiplied by a factor 2. The size of the combined statistical and systematic uncertainty for the sum of the signal and background is indicated by the hatched band. The highest bin in the distributions contains the overflow. The ratio of the data to the sum of the fitted signal and background is shown in the *lower panel* [1]

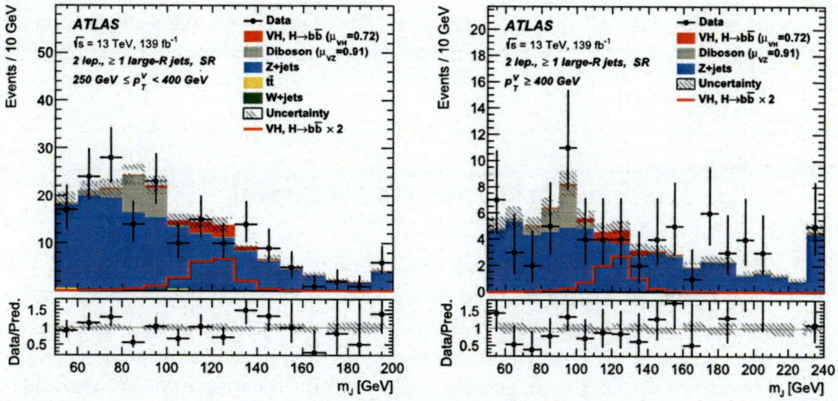

Fig. 8.3 The large-R jet mass post-fit distribution in 2-lepton channel in the SR for 250 GeV$\leq p_{\mathrm{T}}^{V} < 400$ GeV (**a**) and $p_{\mathrm{T}}^{V} \geq 400$ GeV (**b**) regions. The data are shown as *black dots*. The background contributions are shown as filled histograms. The Higgs boson signal with $m_H = 125$ GeV is shown as a filled histogram on the top of the fitted background multiplied by the signal strength μ_{VH}, and unstacked as an unfilled histogram, multiplied by a factor 2. The size of the combined statistical and systematic uncertainty for the sum of the signal and background is indicated by the hatched band. The highest bin in the distributions contains the overflow. The ratio of the data to the sum of the fitted signal and background is shown in the *lower panel* [1]

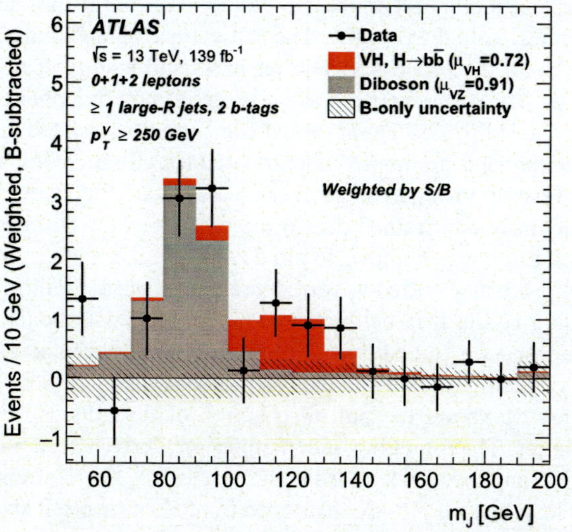

Fig. 8.4 The large-R jet mass m_J distribution in data after subtraction of all the backgrounds except for the WZ and ZZ diboson processes. The contribution of all the three lepton channels and SRs are summed and weighted by their respective values of the ratio of fitted Higgs boson signal with $m_H = 125$ GeV and background yields. The VH signal contribution is multiplied by the measured VH signal strength μ_{VH}. The diboson contribution is normalised to its VZ signal strength μ_{VZ}. The hatched band shows the size of the combined statistical and systematic uncertainties [1]

except for the VZ diboson processes. The plot shows the well pronounced peak around the Z boson mass from the diboson process and the contribution of the Higgs boson.

8.1.3 Nuisance Parameter and Constraints

The pull plots have been scrutinized to point out features or potential issues than need further investigation. The NPs related to systematic uncertainties are grouped by type to help the analyser finding possible issues. This section reports just the pull plots showing the most important aspects. For completeness the other pull plots are shown in Appendix D. In general, pulls without constraints could be just due to small adjustments of the fit and they will be considered of low importance if they are not showing up in the ranking plots. Figure 8.5 shows the pull plots of the NPs obtained from unconditional fit to the Asimov dataset (red points) and to the real dataset (black points). The plots show that the constraints derived from the Asimov dataset match the ones obtained from the real dataset.

Figure 8.5a shows the pull plot of the NPs related large-R uncertainties. As described in Sect. 6.2, the large-R uncertainties can be grouped into resolution uncertainties and scale uncertainties. The resolution uncertainties are further split into jet energy resolution uncertainties, called `FJ_JER` in the plot, and jet mass resolution uncertainties, called `FJ_JMR`. The latter type of uncertainties are treated uncorrelated for all the processes. The jet mass and energy scales are indicated in the plot as `FJ_JMSJES` and they are made up of four components: *baseline*, *modelling*, *tracking* and *statistical*. The systematic uncertainties related to the *baseline* and *modelling* components are treated uncorrelated for all processes. The NPs related to the jet mass resolution and to the jet mass scale (modelling component) of the $t\bar{t}$ process are particularly constrained due to the presence of a characteristic top peak in the mJ distribution which allows to pin down the uncertainties related to the mJ shape. These NPs are also pulled to correct the pre-fit mismodelling of the top peak due to the missing in-situ JMS calibration. The NP related to the jet mass resolution of the diboson process is pulled down because of the disagreement between data and MC simulation near the Z peak in the 0-lepton channel.

Figure 8.5b shows the pull distribution of the NPs associated to the b-tagging uncertainties. The b-tagging uncertainties are derived separately for b-, c-, and light flavour jets and they are labelled in the plot as `FT_B`, `FT_C` and `FT_Light`, respectively. The `FT_B_0` NP is constrained to 65% of its prefit size and pulled down. To understand the NP, an additional fit has been performed decorrelating the NP among samples and analysis regions. The tests show that the pull and the constraint of the NP is originating from the $t\bar{t}$ sample in the CR.

Figure 8.5c shows the pull distribution of the NPs associated to the modelling uncertainties of the $t\bar{t}$ background. The NP related to the $t\bar{t}$ ISR and PS shape uncertainties are indicated in the plot as `TTbarMJISR` and `TTbarMJPS`. The PS NP is constrained for the same reason of the $t\bar{t}$ JMR and JMS NPs. The

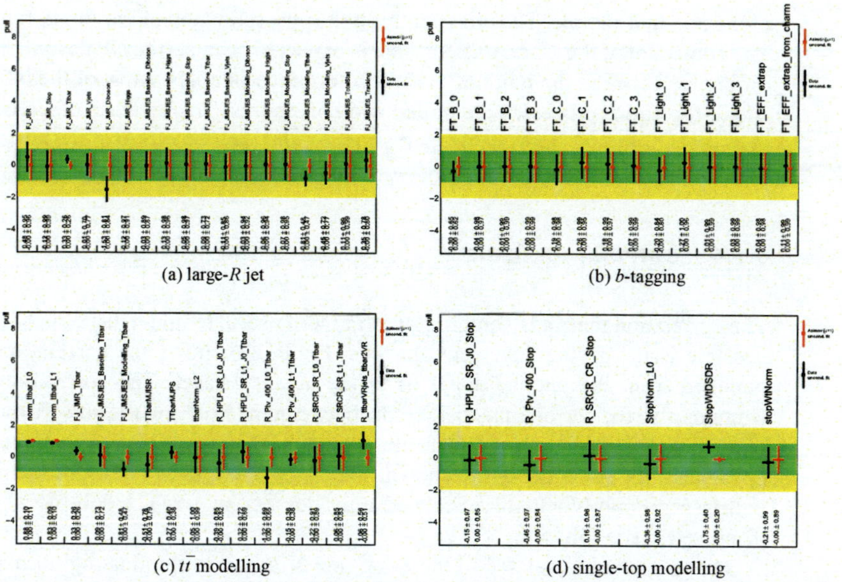

(a) large-R jet (b) b-tagging

(c) $t\bar{t}$ modelling (d) single-top modelling

Fig. 8.5 Detector-related (*top row*) and MC modelling-related (*bottom row*) nuisance parameter pulls, constraints and post-fit normalisation scale factors for the freely floating parameters. All the results are obtained from the unconditional fit in all three lepton channels to the Asimov- (*red*) and the actual dataset (*black*)

`R_Ptv_400_L0_Ttbar` and `R_Ptv_400_L1_Ttbar` NPs control the acceptance between the medium- and high-p_{T}^{V} region in 0- and 1-lepton channels, respectively. The former NP is pulled down because of the under-fluctuation of the data in the high-p_{T}^{V} CR in 0-lepton (Fig. 5.16d). The latter NP is constrained because of the top mismodelling. The `R_ttbarVRjets_ttbar2VR` NP describes the containment of the top decay inside the large-R jet accounting for the ratio between events with two VR track-jets inside the large-R jet and events with three or more VR track-jets inside the large-R jet. The NP is pulled and constrained, and decorrelation tests show that the pull and the constraint are originating from the medium p_{T}^{V} region, as expected. The pull and the constraint are originated from this region because 50% of the events have two VR track-jets inside the large-R jet and the 50% of the events have three or more VR track-jets inside the large-R jet. This means that the two contributions are more distinguishable in the medium p_{T}^{V} region.

Figure 8.5d shows the pull distribution of the NPs associated to the modelling uncertainties of the single-top background. The `StopWtDSDR` NP, which describes the DS-DR variation, is pulled and constrained. The NP is pulled because the fit tends to prefer the prediction of the alternative sample with respect to the nominal sample. This trend is already visible at pre-fit level where there is a better data/MC agreement in the m_J distribution using the alternative single-top Wt sample instead of the nominal sample. The effect of this pull is the reduction of the single-top events. The constraint of the `StopWtDSDR` NP is expected because the NP describes the big difference between the nominal and the alternative samples. Most of the NPs of the single-top and $t\bar{t}$ processes are pulled and constraints. The increase of statistics can help in the future to improve the performance and the modelling of the systematics.

8.1.4 Correlation Matrix

The correlation matrix is another important tool useful to understand the fit and spot possible issues in the statistical analysis. Figure 8.6 shows the correlation matrix obtained from the unconditional fit to the actual dataset. The correlation matrix contains also the correlation coefficients between the PoIs and the NPs. In the matrix only NPs that have at least the absolute value of the correlation of magnitude 0.25 or higher with an other NP are shown. The correlation matrix looks rather diagonal with several sub-blocks of correlated nuisance parameter. The sub-blocks will be discussed in the following.

The first block is related to the b-tagging NPs and to the floating normalization factors. The `FT_C_1` NP is related to the b-tagging uncertainty of c-jets and it is highly correlated with the $t\bar{t}$ floating normalizations `norm_ttbar` in 0- and 1-lepton channels. The correlation is expected because 75% of $t\bar{t}$ events have the large-R jet

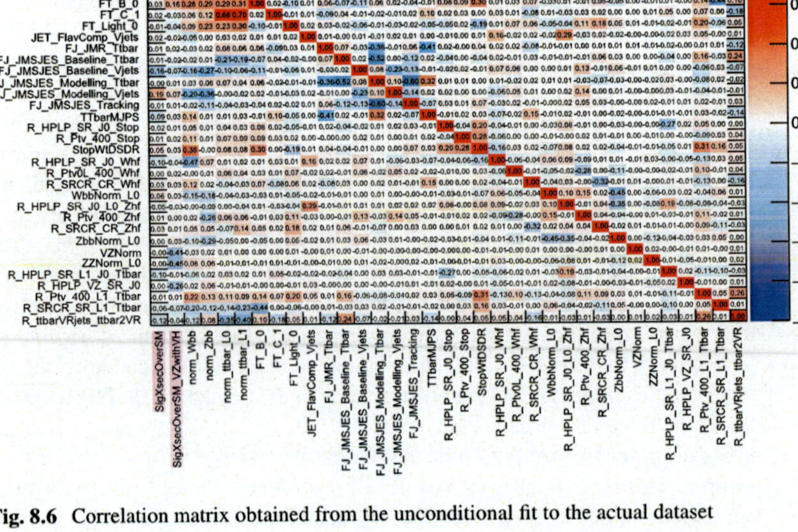

Fig. 8.6 Correlation matrix obtained from the unconditional fit to the actual dataset

composed of one b-labelled jet and one c-labelled jet which is mis-tagged as a b-jet. The FT_B_0 NP related to the b-tagging uncertainty of b-jets shows correlations with the $t\bar{t}$, W+HF (norm_Wbb) and Z+HF (norm_Zbb) floating normalisations. Moreover, the FT_B_0 NP is anti-correlated with the R_SRCR_SR_L1_Ttbar NP related to the extrapolation uncertainty between the SR and CR. The relationship is expected because the extrapolation between the SR and CR depends on the presence of a b-tagged track-jets outside the large-R jet.

The second group of correlated NPs is among the normalisation factors themselves. They are highly correlated and the correlation can be introduced indirectly through the correlation with the b-tagging NPs.

The third block of correlation is made up of NPs related to the large-R uncertainties for the $t\bar{t}$ sample. The FJ_JMR_Ttbar, FJ_JMSJES_Baseline_Ttbar and FJ_JMSJES_Modelling_Ttbar NPs, related to the jet mass resolution and jet mass scale uncertainties, show a high anti-correlation which reflects the fact that all the NPs have the same effect on the top mass peak. These NPs tend to correct the discrepancy between the data and MC events.

The correlation matrix shows also anti-correlation behaviour between WbbNorm_L0 NP, the uncertainty on channel extrapolation for W+HF events in 0-lepton channel, and ZbbNorm_L0, the uncertainty on channel extrapolation for Z+HF events in 0-lepton channel. Anti-correlations between W+ HF and Z+HF events arise from the similarity of the shapes of these two processes in the 0-lepton channel.

Finally, the R_ttbarVRjets_ttbar2VR and R_Ptv_400_L1_Ttbar NPs are correlated. The former NP quantifies to the acceptance uncertainty of the containment of the top decay inside the large-R jet, while the latter NP is related to the extrapolation uncertainty between the medium and high p_T^V region of the $t\bar{t}$ events. The correlation between the two NPs is expected because the containment of the top decay inside the large-R jet varies with p_T^V.

The correlations obtained from the unconditional fit to the data are similar to the ones from the unconditional fit to the Asimov dataset (see Fig. 7.4).

8.1.5 Breakdown of Uncertainties

The effects of the statistical and systematic uncertainties on the measurement of the VH and VZ signal strengths are displayed in Table 8.3. For both signal strength measurements, the uncertainty coming from the statistical component is dominated by the limited size of the dataset, its contribution is two times bigger than the floating normalisation contribution. The systematic contribution can be further split into experimental sources and theoretical and modelling uncertainties. For the VH signal strength, the large-R jets uncertainties have the biggest contribution which is driven by the JMR of the signal process which is the second highest ranked NP (see Fig. 8.7b). Another important contribution to μ_{VH} is from the limited size of the simulated samples followed by the modelling of the signal and of the major back-

Table 8.3 Breakdown of the contribution to the uncertainty on the signal strength μ_{VZ} (second column) and μ_{VH} (third column) obtained from the (1 + 1) PoIs fit to the actual data set. The sum in quadrature of the individual contributions differs from the total uncertainty due to correlations between the nuisance parameters

Source of uncertainty		VZ	VH
Total		0.26	0.37
Statistical		0.15	0.28
↪ Data stat only		0.12	0.25
↪ Floating normalisations		0.06	0.09
Systematic		0.21	0.24
Experimental uncertainties			
Small-R jets		0.02	0.04
Large-R jets		0.09	0.13
E_T^{miss}		0.01	0.01
Leptons		0.01	0.01
b-tagging	b-jets	0.04	0.02
	c-jets	0.01	0.01
	Light-flavour jets	0.01	0.01
	Extrapolation	< 0.01	< 0.01
Pile-up		< 0.01	< 0.01
Luminosity		< 0.01	< 0.01
Theoretical and modelling uncertainties			
Signal		< 0.01	0.04
Backgrounds		0.16	0.10
↪ Z + jets		0.01	0.05
↪ W + jets		0.02	0.06
↪ $t\bar{t}$		0.01	0.04
↪ Single top quark		0.01	0.03
↪ Diboson		0.18	0.03
↪ Multijet		< 0.01	0.01
↪ MC statistical		0.04	0.09

grounds. For the VZ signal strength the major source of uncertainties is from the modelling of the diboson process, followed by the large-R jet contribution.

8.1.6 Nuisance Parameter Ranking

The rankings of the impact of the fifteen most important NPs on the signal strengths are shown in Fig. 8.7. Figure 8.7a shows the impact of the uncertainties on the VZ signal strength. The leading uncertainties for μ_{VZ} are related to the VZ predictions,

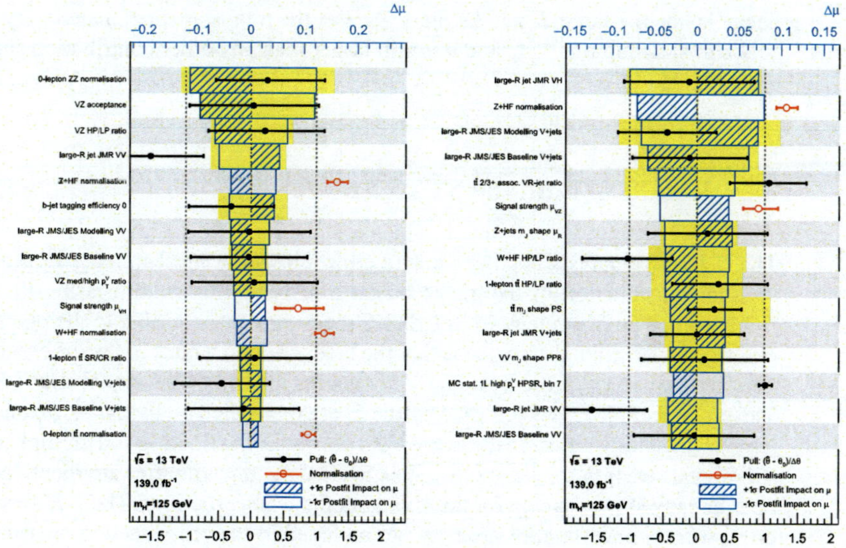

(a) (b)

Fig. 8.7 Impact of systematic uncertainties on the fitted VZ (**a**) and VH (**b**) signal-strength parameters $\hat{\mu}$ sorted in decreasing order. The *boxes* show the variations of $\hat{\mu}$, referring to the top x-axis, when fixing the corresponding individual nuisance parameter to its post-fit value modified upwards or downwards by its post-fit uncertainty, i.e. $\hat{\theta} \pm \sigma_{\hat{\theta}}$, and repeating the fit. The impact of up- and down-variations can be distinguished via the dashed and plane box fillings. The yellow boxes show the pre-fit impact (top x-axis) by varying each nuisance parameter by $\pm 1\sigma$. The *filled circles* show the deviation of the fitted value for each nuisance parameter, $\hat{\theta}$, from their nominal input value θ_0 expressed in standard deviations with respect to their nominal uncertainties $\Delta\theta$ (bottom x-axis). The error bars show the post-fit uncertainties on $\hat{\theta}$ with respect to their nominal uncertainties. The *open circles* show the fitted values and uncertainties of the normalization parameters that are freely floating in the fit

overall acceptance uncertainties and extrapolation uncertainties among additional jet multiplicities. The other highly ranked NP is the large-R jet mass resolution on the diboson. These NPs are highly ranked because they have a high correlation with μ_{VZ}. The highly ranked NPs related to the VZ uncertainties will affect the error on μ_{VZ} and not on the VZ significance because these NPs implement an overall acceptance uncertainty which is degenerate with the μ_{VZ}. The total statistical uncertainty on μ_{VZ} is 0.15 and it is still larger than any other contribution shown in the ranking.

Figure 8.7b shows the impact of the uncertainties on VH signal strength. The leading contribution to the uncertainty is represented by the resolution uncertainty on the large-R jet mass for the VH signal. As before, the contribution of such uncertainty mainly affects the measured μ_{VH}. Other source of uncertainties are the normalisations of the leading Z+jets and diboson backgrounds, the large-R jet mass scale of such processes and acceptance uncertainty for the containment of the top decay inside the large-R jet. As the VZ case, the total statistical uncertainty on the VH signal strength is $^{+0.29}_{-0.28}$ and it is still larger than any other contribution shown in the ranking.

8.2 Results Within the STXS Framework

The $VH(b\bar{b})$ analysis has been extended to provide a measurement of the cross-section as a function of the gauge boson transverse momentum, using the STXS framework. The fit strategy for the STXS measurement is similar to the one already described. The main difference is that the cross-section values in the different STXS bins are the output of the fit to the VH process instead of the signal strength μ_{VH}. The cross-section values returned by the fit are multiplied by the $H \to b\bar{b}$ and $V \to$ leptons branching ratios. In the fit the VZ contribution is floating and multiplied to the VZ signal strength μ_{VZ} as before. The systematic uncertainties are identical to the ones already shown except for the VH systematic uncertainties. The VH theoretical cross-section uncertainties are now not included in the cross-section measurement. Furthermore the split of the STXS bins mitigates the effect of the remaining VH systematics as described in Sect. 1.5.

The STXS bins are defined categorizing the events according to the production mode (WH and ZH) and to the transverse momentum of the vector boson p_{T}^{V} at generator level. There is a cut in p_{T}^{V} at 250 GeV and another one at 400 GeV, in order to be aligned with the ones used at reconstructed level. The adopted splitting is denoted as *main splitting*. As in the *resolved* analysis (see Sect. 1.5), events in the forward Higgs region ($|y_H| > 2.5$) are not included in the measurement. With this categorization scheme, instead of the (1+1) PoIs fit, a (4+1) PoIs fit is performed and five PoIs are extracted: $\sigma^{ZH}_{250-400}$, σ^{ZH}_{400}, $\sigma^{WH}_{250-400}$, σ^{WH}_{400} and μ_{VZ}. Figure 8.8 shows the evolution from the standard signal strength measurement for the VH process to the cross-section measurements performed using the STXS framework. The expected results for the STXS measurements are shown in the following.

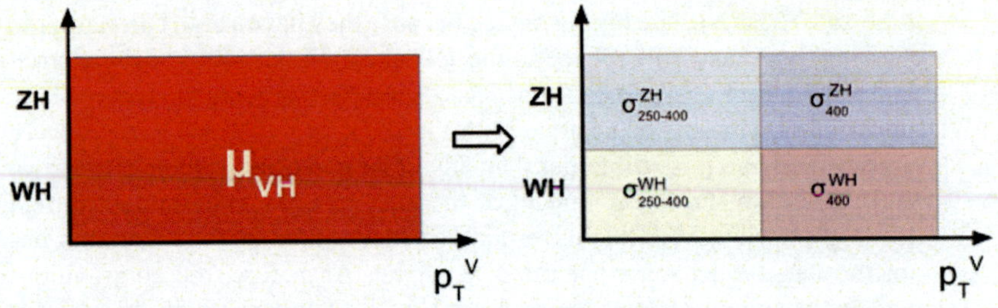

Fig. 8.8 Signal strength measurement of the VH process (*left*) compared to the cross-section measurement (*right*). p_{T}^{V} is the transverse momentum of the V boson at generator level

8.2.1 Comparison Between Truth and Reconstructed Categories

A study of the comparison between the reconstructed signal regions and STXS bins is presented in this section to evaluate the level of matching. The study is done considering the MC signal samples. Figure 8.9a shows the fraction of signal (in percent) for each STXS bin (x-axis) in every reconstructed event category (y-axis), where the HP and LP SRs are merged. The signal fraction in the CRs is not shown in the plot due to the negligible contributions. In the 0-lepton SR, about 20% of the events are from the WH STXS bins. As in the *resolved* analysis, these events reconstructed in the 0-lepton channel are events where the W boson decays leptonically in $\tau + \nu$ and the τ-lepton decays hadronically. The highest level of matching between the truth and the reconstructed categories is achieved in the 2-lepton channel and a quite good matching is obtained in the 1-lepton channel.

Furthermore, the acceptance times efficiency for signal events is studied. The acceptance times efficiency is evaluated as the ratio between the signal events that pass the event selection and the initial number of signal events considering all the Z boson and W boson leptonic decays. Figure 8.9b shows the acceptance times efficiency, in percent, for each STXS bin (x-axis) in every reconstructed event category (y-axis). As before, the HP and LP SRs are merged and the values in the CRs are not reported due to the low contributions. In the 0- and 1-lepton channels the values of the acceptance times efficiency are of the order of 5% in the medium p_{T}^{V} region and of the order of 10% in the high p_{T}^{V} region. In the 2-lepton channel, the values are smaller and of the order 2–4%.

8.2.2 Cross-Section Measurements

A maximum binned likelihood fit is performed using data collected during the Run 2 at $\sqrt{s} = 13$ TeV, corresponding to an integrated luminosity of 139 fb^{-1}. Table 8.4 shows the measured and expected cross-sections of associated production of a Higgs boson, decaying into a bottom quark pairs, with an electroweak boson decaying into leptons. The cross-sections are multiplied by the $H \to b\bar{b}$ and $V \to$ leptons branching ratios. Table 8.4 also reports the statistical and systematic uncertainties on the measured cross-sections in the four STXS bins. All the measurements are in agreement with the SM predictions.

For the VZ process the fitted signal strength μ_{VZ} is:

$$\mu_{VZ} = 0.89^{+0.29}_{-0.23} = 0.89 \pm 0.15(\text{stat.})^{+0.24}_{-0.17}(\text{syst.}) \tag{8.2}$$

in agreement with the result obtained from the (1+1) PoIs fit.

Figure 8.10a shows the measured cross-sections times branching ratios ($\sigma \times B$) normalised to the SM prediction in the four STXS bins. The theoretical uncertainty

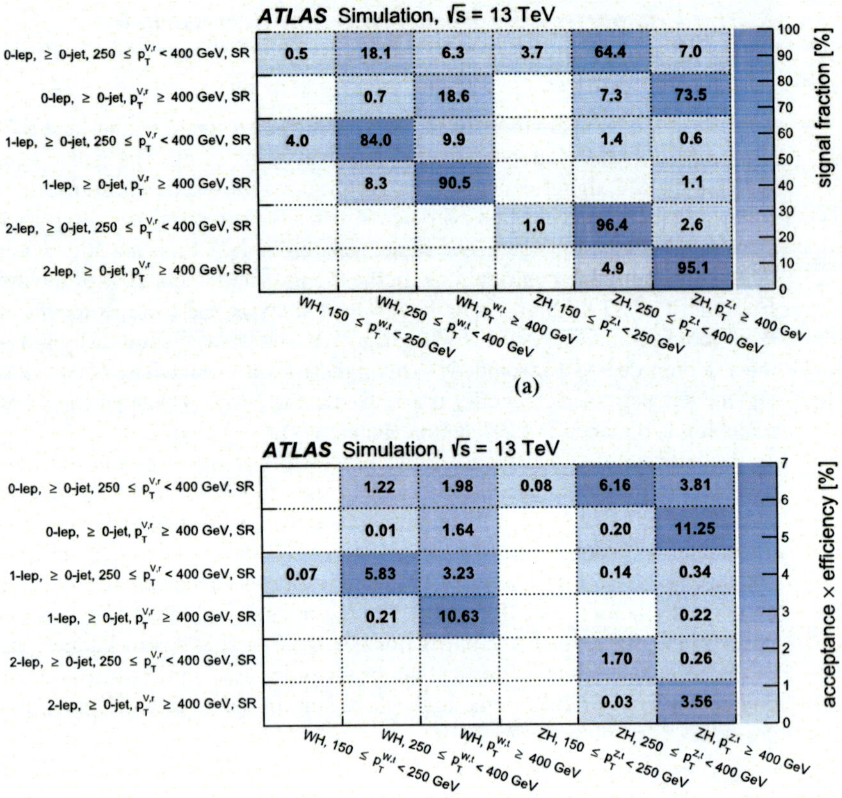

(a)

	WH, 150 ≤ $p_T^{W,t}$ < 250 GeV	WH, 250 ≤ $p_T^{W,t}$ < 400 GeV	WH, $p_T^{W,t}$ ≥ 400 GeV	ZH, 150 ≤ $p_T^{Z,t}$ < 250 GeV	ZH, 250 ≤ $p_T^{Z,t}$ < 400 GeV	ZH, $p_T^{Z,t}$ ≥ 400 GeV
0-lep, ≥ 0-jet, 250 ≤ $p_T^{V,r}$ < 400 GeV, SR	0.5	18.1	6.3	3.7	64.4	7.0
0-lep, ≥ 0-jet, $p_T^{V,r}$ ≥ 400 GeV, SR		0.7	18.6		7.3	73.5
1-lep, ≥ 0-jet, 250 ≤ $p_T^{V,r}$ < 400 GeV, SR	4.0	84.0	9.9		1.4	0.6
1-lep, ≥ 0-jet, $p_T^{V,r}$ ≥ 400 GeV, SR		8.3	90.5			1.1
2-lep, ≥ 0-jet, 250 ≤ $p_T^{V,r}$ < 400 GeV, SR				1.0	96.4	2.6
2-lep, ≥ 0-jet, $p_T^{V,r}$ ≥ 400 GeV, SR					4.9	95.1

(b)

	WH, 150 ≤ $p_T^{W,t}$ < 250 GeV	WH, 250 ≤ $p_T^{W,t}$ < 400 GeV	WH, $p_T^{W,t}$ ≥ 400 GeV	ZH, 150 ≤ $p_T^{Z,t}$ < 250 GeV	ZH, 250 ≤ $p_T^{Z,t}$ < 400 GeV	ZH, $p_T^{Z,t}$ ≥ 400 GeV
0-lep, ≥ 0-jet, 250 ≤ $p_T^{V,r}$ < 400 GeV, SR		1.22	1.98	0.08	6.16	3.81
0-lep, ≥ 0-jet, $p_T^{V,r}$ ≥ 400 GeV, SR		0.01	1.64		0.20	11.25
1-lep, ≥ 0-jet, 250 ≤ $p_T^{V,r}$ < 400 GeV, SR	0.07	5.83	3.23		0.14	0.34
1-lep, ≥ 0-jet, $p_T^{V,r}$ ≥ 400 GeV, SR		0.21	10.63			0.22
2-lep, ≥ 0-jet, 250 ≤ $p_T^{V,r}$ < 400 GeV, SR					1.70	0.26
2-lep, ≥ 0-jet, $p_T^{V,r}$ ≥ 400 GeV, SR					0.03	3.56

Fig. 8.9 **a** Fraction of signal events and **b** acceptance times efficiency from each STXS bin (x-axis) in every reconstructed event category (y-axis). The values are in percent. The HP and LP SRs in 0- and 1-lepton channels have been merged. The signal contribution in the CRs is negligible and therefore not reported in the plots. Entries with acceptance times efficiency below 0.01% or signal fractions below 0.1% are not shown [1]

on the normalised cross-section times branching ratios is shown as a grey area. The main contribution to the uncertainty is statistical, which becomes larger in the high p_T^V region. The ZH measurements have a bigger theoretical uncertainty on the cross-sections than the WH measurements because of the limited precision of the $gg \to ZH$ samples. The results extend those obtained by the *resolved* analysis [2], which provides a more precise measurement of the cross-section because of the sophisticated multivariate analysis techniques adopted and because of the possibility to use better calibrated physics objects. However, the *resolved* analysis does not provide cross-section measurements for $p_T^V \geq 400$ GeV.

Table 8.4 Measured and predicted cross-sections times BRs with corresponding uncertainties in the four STXS bins [1]

Measurement region ($\lvert y_H \rvert < 2.5, H \to b\bar{b}$)	SM prediction [fb]	Result [fb]	Stat. unc. [fb]	Syst. unc. [fb]
$H(\to b\bar{b})W(\to \ell\nu);\ p_T^W \in [250, 400[$ GeV	5.83 ± 0.26	$3.33^{+4.78}_{-4.56}$	$^{+3.56}_{-3.44}$	$^{+3.19}_{-2.99}$
$H(\to b\bar{b})W(\to \ell\nu);\ p_T^W \in [400, \infty[$ GeV	1.25 ± 0.06	$2.10^{+1.18}_{-1.07}$	$^{+1.02}_{-0.95}$	$^{+0.59}_{-0.49}$
$H(\to b\bar{b})Z(\to \ell\ell, \nu\nu);\ p_T^Z \in [250, 400[$ GeV	4.12 ± 0.45	$1.40^{+3.12}_{-2.89}$	$^{+2.44}_{-2.35}$	$^{+1.94}_{-1.68}$
$H(\to b\bar{b})Z(\to \ell\ell, \nu\nu);\ p_T^Z \in [400, \infty$ GeV	0.72 ± 0.05	$0.17^{+0.67}_{-0.63}$	$^{+0.58}_{-0.53}$	$^{+0.33}_{-0.34}$

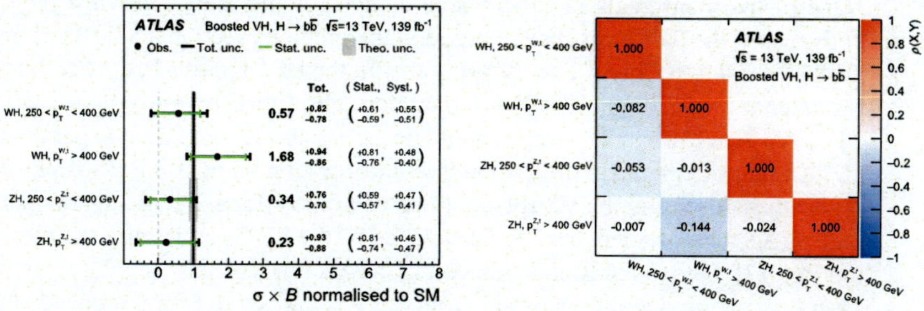

(a) (b)

Fig. 8.10 **a** Measured cross-section times BRs normalised to the SM predictions. The *grey* error bands correspond to the theoretical uncertainty on the normalised $\sigma \times$BR. **b** Observed correlations among the cross-sections in the STXS bins. The colour indicates the size of the correlation [1]

In each STXS bin, the principal sources of systematic uncertainties are similar to those affecting the VH signal strength. The leading source is from the large-R uncertainties, followed by the background modelling uncertainties. Another important contribution is from the limited size of the simulated samples.

The expected and measured significances for the measurements in the four STXS bins are reported in Table 8.5. Table 8.5 also shows the expected and observed significance for the VZ process. The VH observed significance varies between 0.3σ and 2σ. Small anti-correlations of the order of 5–10% are observed among the cross-sections in different STXS bins (Fig. 8.10b).

Table 8.5 Expected and observed significance for the measured cross-sections in the four STXS bins and for the VZ process [1]

Measurement	Expected sig.	Observed sig.
$WH[250, 400[$ GeV	1.26	0.73
$WH[400, \infty[$ GeV	1.27	2.02
$ZH[250, 400[$ GeV	1.38	0.49
$ZH[400, \infty[$ GeV	1.12	0.26
VZ	5.56	5.20

8.2.3 Effective Field Theory Interpretations

One of the main goals of the *boosted* analysis is the search of BSM physics. For this reason the STXS results presented in the previous section are further interpreted using an Effective Field Theory approach. In the EFT approach, the SM Lagrangian is extended with dimension D=6 operators multiplied by the Wilson coefficients. The Wilson coefficients, which should be zero in the SM, are used to parametrise the STXS measurements in order to extract information on possible BSM physics. The procedure for fitting the Wilson coefficients of the effective Lagrangian operators to the STXS measurement has been developed by the LHC community and documented in Ref. [3]. Due to the limited number of STXS bins, not all the Wilson coefficients can be measured simultaneously. The study is carried out on a reduced set of five Wilson coefficients, which is the same set of the *resolved* analysis as documented in Sect. 1.6.1. The set is composed of four Wilson coefficients, c_{HWB}, c_{HW}, c_{Hu}, c_{Hq3}, to which the analysis is most sensitive, and the $|c_{dH}|$ coefficient which directly affects the $H \to b\bar{b}$ decay width.

As reported in Sect. 1.6, the parametrisation of the STXS takes into account linear terms, originating from the interference between the SM and the non-SM amplitudes, and a quadratic terms, originating from the squared non-SM amplitudes. The constraints on the five coefficients are extracted from the one-dimensional fit where only a single parameter is varied while the remaining four are set to zero. Figure 8.11 shows the negative log-likelihood profile as a function of the single Wilson coefficient. The fits are performed using a linear parametrisation (blue lines) and using a parametrisation that includes both linear and quadratic terms (orange lines). The analysis has the best sensitivity to c_{Hq3} Wilson coefficient which is constrained at 68% CL to be no more than few percent. Differently the constraints on the c_{HW} and c_{Hu} coefficients vary from 15–40%, and c_{HWB} and $|c_{dH}|$ coefficients have much weaker constraints. At 95% CL considering the linear and quadratic parametrisation, the observed limits on c_{Hq3} are more stringent in the *boosted* analysis with respect to the *resolved* analysis. This means that the *boosted* analysis provides a tighter constraint on the EFT parameter. An opposite trend is observed for the other Wilson coefficients. The results of the fit are shown in Fig. 8.12 and summarized in Table 8.6.

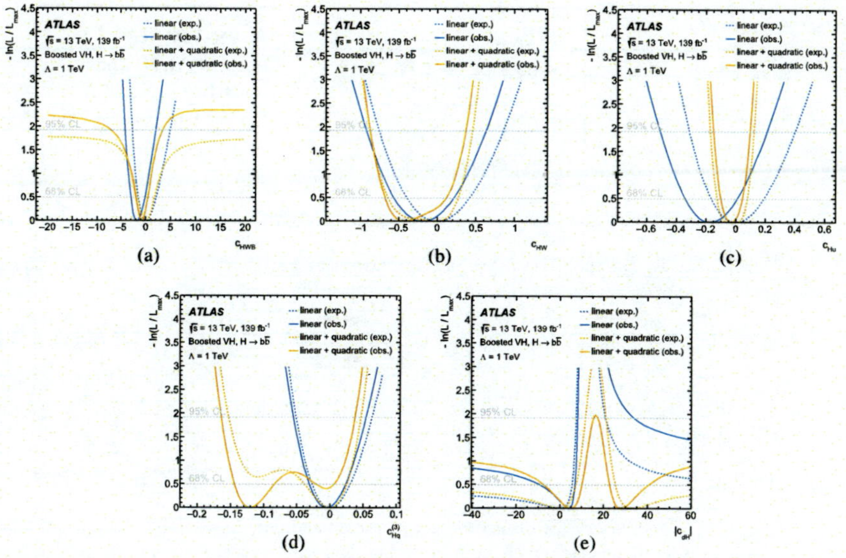

Fig. 8.11 Observed (*solid lines*) and expected (*dotted lines*) profiled negative log-likelihood functions for one-dimensional fits to constrain a single coefficient of an effective Lagrangian when the other coefficients are assumed to vanish. The coefficients c_{HWB}, c_{HW}, c_{Hu}, c_{Hq3} and $|c_{dH}|$ are shown for the case where only linear (*blue lines*) or linear and quadratic (*orange lines*) terms are considered [1]

Two-dimensional fits have been performed for all the combinations of the four Wilson coefficients to which the analysis has the greatest sensitivity. In the fit, only the scanned Wilson coefficients are considered, the remaining two are fixed to zero. Figure 8.13 shows the observed confidence interval at 68% (dashed lines) and 95% CLs on the pair of c_{Hq3} and c_{Hu} Wilson coefficients. The limits are obtained considering the linear parametrisation (blue lines) and including the quadratic terms (orange lines) of the VH cross-section times and branching ratios. The limits extracted considering the linear only and linear plus quadratic parametrisations are different, underlying the importance of the quadratic terms.

The main goal of the *boosted* analysis is to explore the high-p_{T}^{V} phase space region in the VH production extending the STXS scheme with a cut on p_{T}^{V} at 400 GeV. To quantify the benefits arising from this improved granularity, the limits on the Wilson coefficients obtained from the default analysis scheme with 4 STXS bins have been compared with the limits obtained considering only 2 STXS bins without the cut at $p_{\mathrm{T}}^{V} = 400$ GeV. The cut at $p_{\mathrm{T}}^{V} = 400$ GeV brings 70% improvement on the confidence intervals extracted from the one-dimensional and two dimensional fits.

The effects of the BSM physics have been studied using also the eigenvector method as in the *resolved* analysis (see Appendix A). The idea of the method is to change operator basis with new eigenvectors whose eigenvalues are linear combinations of the four Wilson coefficients mostly affecting the analysis. The main differ-

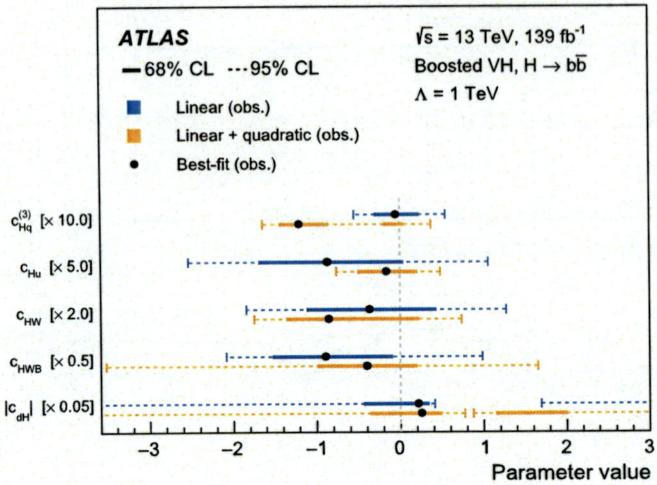

Fig. 8.12 Summary of the observed best-fit values and one-dimensional confidence intervals for the Wilson coefficients to which the analysis has the greatest sensitivity and for the $|c_{dH}|$ coefficient which directly affects the $H \to b\bar{b}$ decay width. Limits are shown for the case where only linear (*blue*) and linear and quadratic (*orange*) terms are considered. Confidence intervals are shown at both 68% CL (*solid line*) and 95% CL (*dashed line*) [1]

Table 8.6 Observed one-dimensional confidence intervals for the Wilson coefficients to which this analysis has the greatest sensitivity and the $|c_{dH}|$ coefficient which directly affects the $H \to b\bar{b}$ decay width. Confidence intervals are shown at both 68% CL and 95% CL. Numbers are shown for the linear-only and linear+quadratic parametrisations [1]

Coefficient	Linear	Linear + quadratic	Linear	Liner + quadratic
	68% CL		95%	
c_{Hq3}	[−0.031, 0.024]	[−0.14, −0.084] ∪ [−0.020, 0.006]	[−0.055, 0.055]	[−0.16, 0.038]
c_{Hu}	[−0.34, 0.0094]	[−0.1, 0.043]	[−0.51, 0.21]	[−0.15, 0.097]
c_{HW}	[−0.56, 0.22]	[−0.68, 0.12]	[−0.92, 0.64]	[−0.92, 0.64]
c_{HWB}	[−3.1, −0.17]	[−2, 0.41]	[−4.2, 2]	[−7.1, 3.3]
$\|c_{dH}\|$	[−8.9, 7]	[−7.2, 10] ∪ [23, 40]	[−∞, 8.2] ∪ [34, ∞]	[−∞, 16] ∪ [18, ∞]

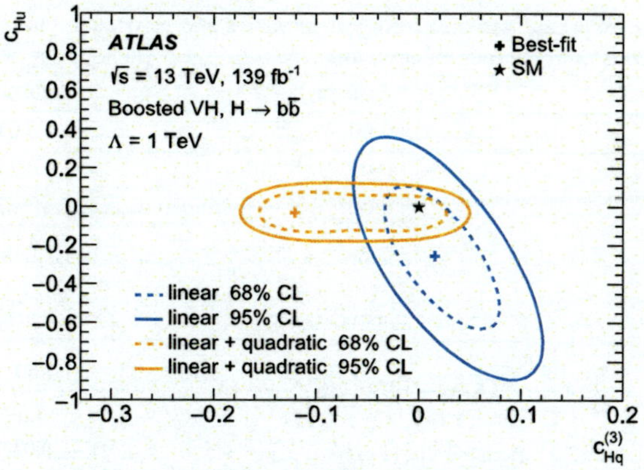

Fig. 8.13 Observed confidence interval at 68% (*dashed lines*) and 95% (*solid lines*) CLs on the pair of c_{Hq3} and c_{Hu} Wilson coefficients. The limits are shown for the case where only linear (*blue lines*) and linear and quadratic (*orange lines*) terms are considered. The best-fit points are marked by a *cross*. The SM prediction is marked by a star [1]

Table 8.7 Linear combination of the Wilson coefficients used to define the eigenvectors. The Wilson coefficients with eigenvalues less than 0.10 have been omitted in the table for better readability. All the modifications that alter the branching ratio are absorbed into an additional independent term $\mathcal{I}_{BR}$ [1]

Eigenvector name	Eigenvector definition
c_{E0}	$0.99 \cdot c_{Hq3} + 0.11 \cdot c_{Hu}$
c_{E1}	$0.82 \cdot c_{Hu} - 0.49 \cdot c_{Hq1} - 0.24 \cdot c_{Hd} - 0.13 \cdot c_{Hq3}$
c_{E2}	$0.67 \cdot \mathcal{I}_{BR} + 0.66 \cdot c_{HW} + 0.18 \cdot c_{Hq1} - 0.16 \cdot c_{Hl3} + 0.14 \cdot c_{HWB} + 0.12 \cdot c_{ll1}$
c_{E3}	$0.70 \cdot c_{Hq1} + 0.52 \cdot c_{HWB} + 0.27 \cdot c_{Hu} - 0.27 \cdot c_{HW} - 0.24 \cdot c_{Hd} + 0.13 \cdot c_{HB}$

ence between this approach and the previous one is that in the former a simultaneous fit to all the four eigenvectors is performed, while in the latter the fit is performed assuming that all the coefficients except one or two are zero. The definition of the new four Wilson coefficients is reported in Table 8.7, considering a linear parametrisation and an additional independent parameter $\mathcal{I}_{BR}$ for the branching ratio $H \to b\bar{b}$. In Table 8.7 the eigenvectors are ordered in term of experimental sensitivity, the c_{E0} eigenvector has the greatest sensitivity and c_{E3} eigenvector has the least sensitivity. The leading c_{E0} eigenvector is dominated by c_{Hq3} Wilson coefficient which is the most constrained coefficient by the analysis. The c_{E1} eigenvector is dominated by c_{Hu} Wilson coefficient with a sizeable contribution of c_{Hq1} and c_{Hd} Wilson coefficients. The c_{E2} eigenvector is a combination of the branching ratio parameter $\mathcal{I}_{BR}$ and c_{HW}, while the c_{E3} eigenvector is a combination of c_{Hq1} and c_{HWB} Wilson coefficients.

Table 8.8 Expected and observed best-fit values and associated uncertainties (68% CL) from a simultaneous fit of the four eigenvectors [1]

Eigenvector	Expected	Observed
c_{E0}	$0.000^{+0.030}_{-0.027}$	$-0.010^{+0.027}_{-0.025}$
c_{E1}	$0.00^{+0.20}_{-0.19}$	$-0.21^{+0.19}_{-0.20}$
c_{E2}	$0.00^{+0.71}_{-0.67}$	$-0.62^{+0.70}_{-0.66}$
c_{E3}	$0.0^{+2.8}_{-2.7}$	$0.4^{+2.8}_{-2.7}$

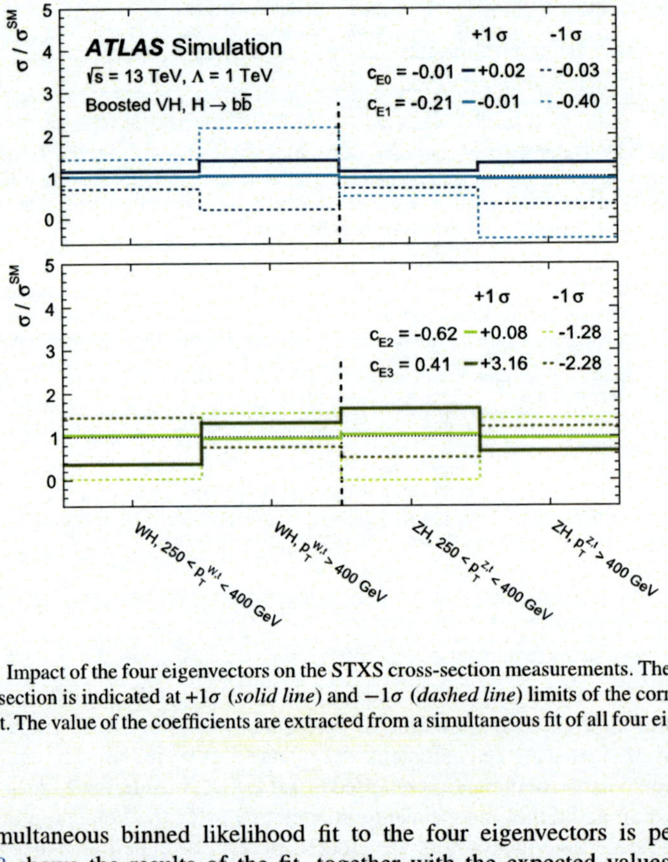

Fig. 8.14 Impact of the four eigenvectors on the STXS cross-section measurements. The change of the cross-section is indicated at $+1\sigma$ (*solid line*) and -1σ (*dashed line*) limits of the corresponding coefficient. The value of the coefficients are extracted from a simultaneous fit of all four eigenvectors [1]

A simultaneous binned likelihood fit to the four eigenvectors is performed. Table 8.8 shows the results of the fit, together with the expected values. All the eigenvectors are compatible with zero. The best-fit values are used to evaluate the impact of the variations of the four eigenvectors to the expected cross-sections as shown in Fig. 8.14. The c_{E0} and c_{E1} coefficients have a large impact on the high-p_T^V bins, while the c_{E2} and c_{E3} coefficients have a large impact in the medium-p_T^V bins.

8.3 What's Next?

Using the full Run 2 dataset, both $VH(b\bar{b})$ *resolved* and *boosted* analyses provided VH cross-section measurements and limits on possible BSM effects. The next natural step will be to combine the information of the two analyses. The phase space of the two analyses significantly overlaps and the combination is not straightforward. A first step to draft a combination strategy is to study the fraction of signal events reconstructed by both analyses. These events represent the *overlap events*. Figure 8.15 shows the fractions of simulated VH signal events, as a function of the p_T^V, reconstructed only by the *resolved* or *boosted* analysis together with the overlap events in the 0-lepton channel. The sum of the three curves is equal to the unity. The fraction of events reconstructed only by the *resolved* analysis starts to decrease at high energy. This trend is expected because the reconstruction of the Higgs boson decay products based on the physics objects used in the *resolved* analysis becomes inefficient at high p_T^V. The crossing point between the *resolved* and *boosted* curves is at $p_T^V = 300$ GeV meaning that above this point the fraction of the events reconstructed by the *boosted* analysis is larger than the fraction of events selected only by the *resolved* analysis. In addition, for what concerns the signal, 60% of the selected events in the *boosted* analysis are also selected by the corresponding *resolved* analysis. In the 250 GeV $\leq p_T^V < 400$ GeV region, the *boosted* analysis is able to select

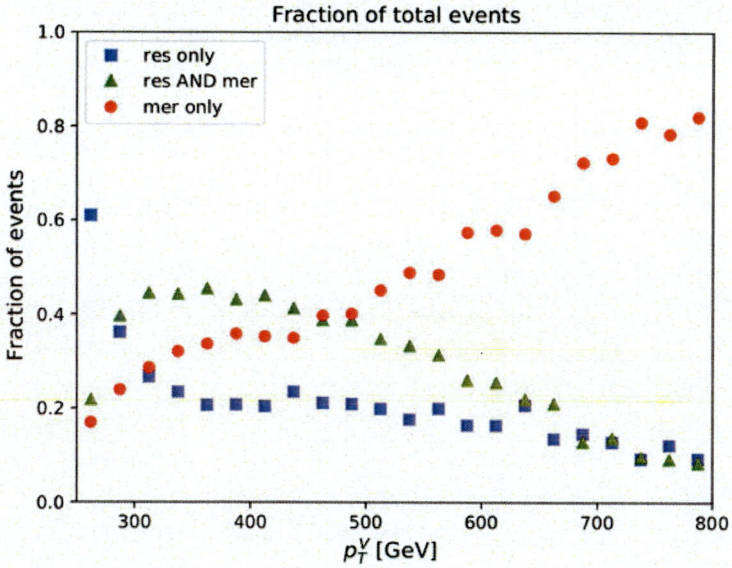

Fig. 8.15 Fraction of the simulated signal VH events reconstructed only by the *resolved* (*blue squares*) and *boosted* (*red dots*) analysis, together with the fraction of overlap events (*green triangles*) as a function of the p_T^V in the 0-lepton channel

35% more events with respect to the *resolved* analysis, while in the high-p_T^V region ($p_T^V \geq 400$ GeV) the *boosted* analysis reconstructs 75% more events than the *resolved* analysis. For a complete study of the phase space overlap, the same plot should be done considering also the background events.

A huge effort will be on the study of the different strategies to combine the two analysis because there are several possibilities. The easiest solution is to reconstruct the events with $p_T^V < 400$ GeV using the *resolved* analysis and reconstruct events with $p_T^V \geq 400$ GeV using the *boosted* analysis. This solution has the disadvantage to loose 20% signal events in the high-p_T^V region selected only by the *resolved* analysis. To avoid signal loss, all the events selected by any of the two analyses could be used, by, for instance, giving priority to the *resolved* selection, and testing the remaining events with the *boosted* selection. This implementation is more complex with respect to the first one but it gives the possibility to avoid signal loss.

References

1. Aad et al G (2020) Measurement of the associated production of a Higgs boson decaying into b-quarks with a vector boson at high transverse momentum in pp collisions at p s = 13 TeV with the ATLAS detector. arXiv: 2008.02508 [hep-ex]
2. Aad et al G (2020) Measurements of WH and ZH production in the H → $b\bar{b}$ decay channel in pp collisions at 13 TeV with the ATLAS detector. arXiv: 2007.02873 [hep-ex]
3. Hays C, Sanz Gonzalez V, Zemaityte G (2019) Constraining EFT parameters using simplified template cross sections. Tech. rep. LHCHXSWG-2019-004, CERN. https://cds.cern.ch/record/2673969

Chapter 9
Cross-Section Measurements of the VZ Production

The chapter describes a spin-off of the $VH(b\bar{b})$ boosted analysis which performs cross-section measurements of the VZ process. The analysis is new and not completed but preliminary results are shown in the section. For the first time, $ZZ(b\bar{b})$ and $WZ(b\bar{b})$ cross-sections at $\sqrt{s} = 13$ TeV are measured in two p_{T}^{V} regions.

9.1 Motivation and VZ Composition

In the $VH(b\bar{b})$ *boosted* analysis the default fit is (1+1) PoIs fit where the VH and the VZ contributions are extracted simultaneously. The VH and VZ processes have a similar topologies and the simultaneous fit allows to test the analysis on an irreducible background. For this reason the VZ results is used a cross-check of the VH analysis.

Some tests have been performed to evaluate the sensitivity of the analysis on the VH and VZ processes. The tests show that the analysis has a better sensitivity to the diboson process than to the Higgs production mode, in particular the precision on the ZZ process is twice better with respect to the precision on the ZH process. The idea of this spin-off analysis is to re-use the $VH(b\bar{b})$ *boosted* analysis to perform cross-section measurements of the VZ process.

The study of the WZ and ZZ production modes in *p*-*p* collisions provides an important test of the gauge boson sector of the SM. The CMS Collaboration already presented a measurement of the VZ cross-section in the $VZ \to Vb\bar{b}$ decay modes where the V boson decays leptonically re-using the $VH(b\bar{b})$ analysis [1]. In this analysis the cross-sections are measured simultaneously for the WZ and ZZ production with $p_{\mathrm{T}}^{V} > 100$ GeV using a dataset corresponding to an integrated luminosity of 18.9 fb^{-1} with $\sqrt{s} = 8$ TeV.

The analysis presented here is an unprecedented attempt to measure $VZ(b\bar{b})$ cross-sections at $\sqrt{s} = 13$ TeV, in the high p_{T}^{V} regime. As already mentioned in Sect. 3.4.4, the VZ process includes the WZ and ZZ processes. The ZZ production

G. Di Gregorio, *A Beauty-ful Boson*, Springer Theses,
https://doi.org/10.1007/978-3-031-20013-7_9

Table 9.1 Monte Carlo VZ fraction in the analysis summing the contributions of the three lepton channels and all the analysis regions. The asterisk * indicates the events in which the hadronic decay of Z boson is not b-quarks

$Z(\nu\bar{\nu}/l\bar{l})Z(b\bar{b})$	$W(l\nu)Z(b\bar{b})$	$Z(\nu\bar{\nu}/l\bar{l})Z(q\bar{q})[*]$	$W(l\nu)Z(q\bar{q})[*]$	$W(qq')Z(\nu\bar{\nu}/l\bar{l})$
47.7%	43.8%	1.9%	3.3%	3.3%

mode is split into two contributions: $qq \to ZZ$ and $gg \to ZZ$. The ZZ events that contribute to the analysis are events in which one Z boson decays leptonically and the other Z boson decay hadronically. For the WZ events the main contribution in the analysis is from events in which the W boson decays leptonically and the Z boson decays hadronically. The major input is from events in which the Z boson decays into a $b\bar{b}$ pair. There is also a small contribution from events in which the W boson decays hadronically and the Z boson decays leptonically. Table 9.1 briefly summarizes the different fractions of the MC VZ contributions in the $VH(b\bar{b})$ analysis. About 90% of the VZ events in the analysis are $VZ(b\bar{b})$ events in which the Z boson decays into a pair of b-quarks and the V boson decays leptonically. Almost 3.3% of the events are $W(qq')Z(\nu\nu/ll)$ events, in which the vector boson that decays hadronically is a W boson instead of the Z boson. The remaining 5.2% of the VZ events are $VZ(q\bar{q})$ events in which the Z boson decays hadronically but not into a $b\bar{b}$ pair.

9.2 VZ Simplified Template Cross-Section Bins

A STXS-like framework is used to perform the cross-section measurement for the VZ process. The use of the STXS framework for the VZ mass was never use before and it per se a challenge. The definition of the VZ STXS bins replicates the one already discussed for the VH process. Each bin is defined using the generator values of the measured quantities. Only the VZ events in which the Z boson decays hadronically and the V boson decays leptonically are considered as signal events. The $W(qq')Z(\nu\bar{\nu}/l\bar{l})$ events are considered as background events of the analysis. The VZ events are categorized according to the production and decay modes in order to have five categories: $gg \to Z(q\bar{q})Z(\nu\bar{\nu})$, $qq \to Z(q\bar{q})Z(\nu\bar{\nu})$, $qq \to Z(q\bar{q})W(l\nu)$, $gg \to Z(q\bar{q})Z(l\bar{l})$ and $qq \to Z(q\bar{q})Z(l\bar{l})$. Each category is divided according to the transverse momentum of the vector boson which decays leptonically. The events are split applying four cuts at p_T^V = 75 GeV, 150 GeV, 250 GeV and 400 GeV. This configuration with 25 bins is called *maximum splitting* and it is shown in Fig. 9.1a. The measurement presented in this thesis uses a coarser splitting referred as *main splitting*. The different ZZ productions and Z leptonic decay modes are all merged together. All the VZ events with $p_T^V < 250$ GeV are fixed to their SM values in the fit. Figure 9.1b shows the VZ STXS categorization in the *main splitting* scheme and the four STXS bins where the cross-section measurements have been performed.

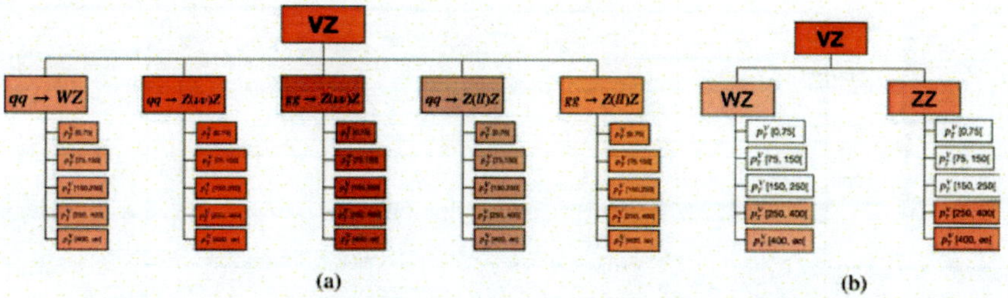

Fig. 9.1 **a** *Maximum splitting* and **b** *main splitting* of the VZ events in STXS bins. The boundaries of the truth p_{T}^{V} regions are expressed in GeV. In the *main splitting* the measurements are performed only in the bin shown in *red*, the bins shown in *white* are fixed to their SM values

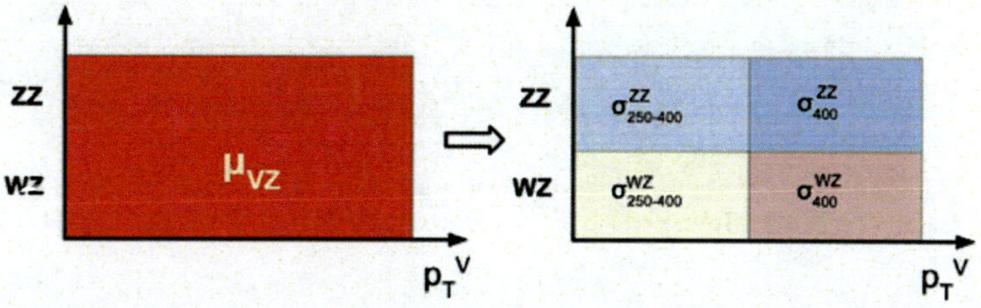

Fig. 9.2 Pictorial representation of the PoIs for the signal strength measurement of the VZ process (*left*) compared to the cross-section measurements (*right*)

The goal of this new analysis is achieved by replacing the VZ signal strength μ_{VZ} with the cross-section measurements in the four STXS bins (see Fig. 9.2) performing a (4+4) PoIs fit. The eight PoIs extracted from the fit are: $\sigma^{ZH}_{250-400}$, σ^{ZH}_{400}, $\sigma^{WH}_{250-400}$, σ^{WH}_{400}, $\sigma^{ZZ}_{250-400}$, σ^{ZZ}_{400}, $\sigma^{WZ}_{250-400}$, σ^{WZ}_{400}.

Some modifications to the framework used for the VH analysis are required. The measurement of the cross-sections needs to split the VZ simulated events and to update the VZ systematics uncertainties. The overall theoretical uncertainties should be removed and residual uncertainties need to be considered while performing a cross-section measurement. The VZ analysis is at the early stage and the systematic uncertainties have been not yet changed, meaning that the VZ uncertainties are over-conservative. Only the p_{T}^{V} acceptance uncertainty on the VZ events is neglected since the cross-section measurements have been performed in the two p_{T}^{V} regions independently. The theoretical uncertainties on the predicted cross-sections have been specifically derived for the VZ measurement following the recommendations available in literature and they are presented in Sect. 9.3. To validate the split of the VZ inputs, a merging procedure of the VZ events has been applied and the (1+1) PoIs fit has been performed as a cross check.

The expected cross-sections times the branching ratios of the $VZ(b\bar{b})$ events in each bin of the *maximum splitting* scheme were not available in literature and they have been evaluated for this analysis. Figure 9.3 shows the obtained cross-sections times branching ratios of the $VZ(b\bar{b})$ events in the *maximum splitting* scheme. The

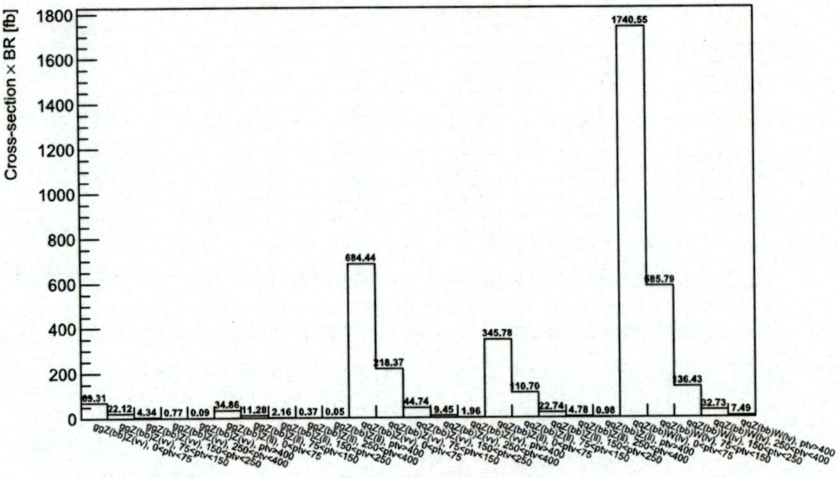

Fig. 9.3 SM $VZ(b\bar{b})$ cross-sections times branching ratios in each bin of the *maximum splitting* scheme at $\sqrt{s} = 13$ TeV

considered branching ratios are the branching ratio of the leptonic decay of the V boson and the branching ratio of the decay of the Z boson into $b\bar{b}$ pair since more than 90% of the VZ events selected in the analysis have a Z boson which decays into b-quarks. The cross-sections times branching ratios are obtained by computing from simulated VZ samples the fractions of events in each bin, and multiplying such fractions by the theoretical predictions of the inclusive cross-section. The theoretical prediction of the inclusive cross-section are calculated at NLO in QCD (see Sect. 3.4.4). The qq-initiated MC samples have been generated at LO and NLO in QCD depending on the parton multiplicity, while the gg-initiated MC samples have been generated at LO in QCD.

The expected $VZ(b\bar{b})$ cross-sections times the branching ratios in each bin of the *main splitting* are shown in Fig. 9.4.

9.2.1 Comparison to Reconstructed Categories

The expected VZ event yield in each reconstructed event category (y-label) originating from the STXS bin (x-label) is shown in Fig. 9.5a. The number of reconstructed events with a truth $p_T^V < 250$ GeV is low because the analysis selects only events with $p_T^V \geq 250$ GeV. The events in these regions are fixed in the fit to their SM values because the statistics is not enough to perform a meaningful measurement.

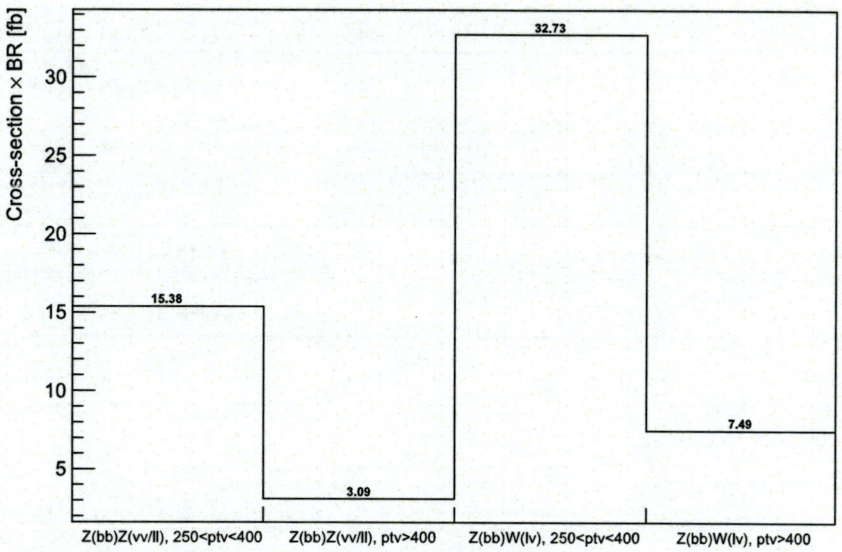

Fig. 9.4 SM $VZ(b\bar{b})$ cross-sections times branching ratios in each bin of the *main splitting* scheme at $\sqrt{s} = 13$ TeV

The fraction of VZ events in each reconstructed event category (y-label) originating from the STXS bins (x-label) is shown Fig. 9.5b. The 2-lepton channel shows a very good correspondence between the truth and the reconstructed categories with a matching of 95%, followed by the 1-lepton channel with a matching of 80–90%. Differently in the 0-lepton channel there is a 20% contribution from the WZ events. The same trend has been observed considering VH signal process in which 20% of the events reconstructed in the 0-lepton channel are coming from the WH process. The WZ events reconstructed in the 0-lepton channel are events where the W boson decays in $\tau + \nu$ and then the τ lepton decays hadronically.

9.3 Theoretical Cross-Sections Uncertainties

The theoretical uncertainties on the cross-sections, as the theoretical predictions on the cross-sections, were not available in literature and they have been derived for this thesis. These uncertainties take into account the missing higher-order terms in the QCD expansion and uncertainties induced by the choices of the parton distribution functions (PDFs) and α_S. The uncertainties have been derived specifically for each source of uncertainties and then summed in quadrature to obtain the total uncertainty on the theoretical prediction. A brief description of the different components of the theoretical cross-sections uncertainties is reported in the following sub-sections.

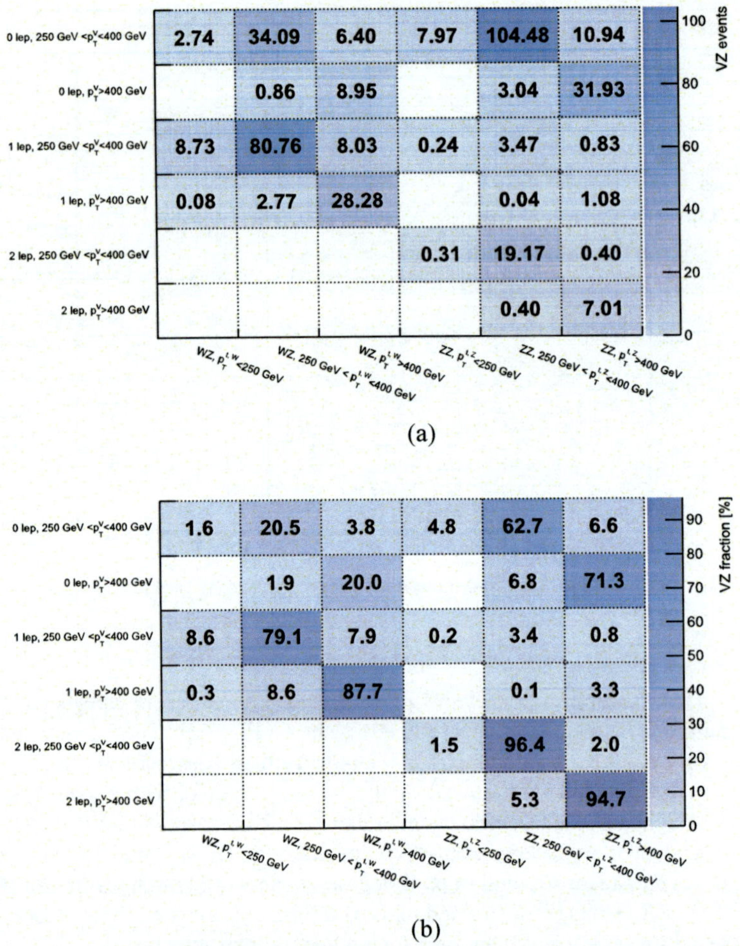

Fig. 9.5 **a** VZ events and **b** VZ fraction from each STXS bin (x-axis) in every reconstructed event category (y-axis). The HP and LP SRs in 0- and 1-lepton channels have been merged and the contribution in the CRs is not shown because it is negligible

9.3.1 QCD Scale Uncertainties

The QCD scale uncertainties account for the impact of the higher order terms in the expansion of the cross-section that have been neglected. The scale variations are evaluated in the inclusive p_T^V bins because,due to cancellations, in the exclusive p_T^V bins the variations could underestimate the uncertainties. The scale variations in the inclusive p_T^V bins are then propagated to the exclusive p_T^V bins using the Stewart-Tackmann method [2].

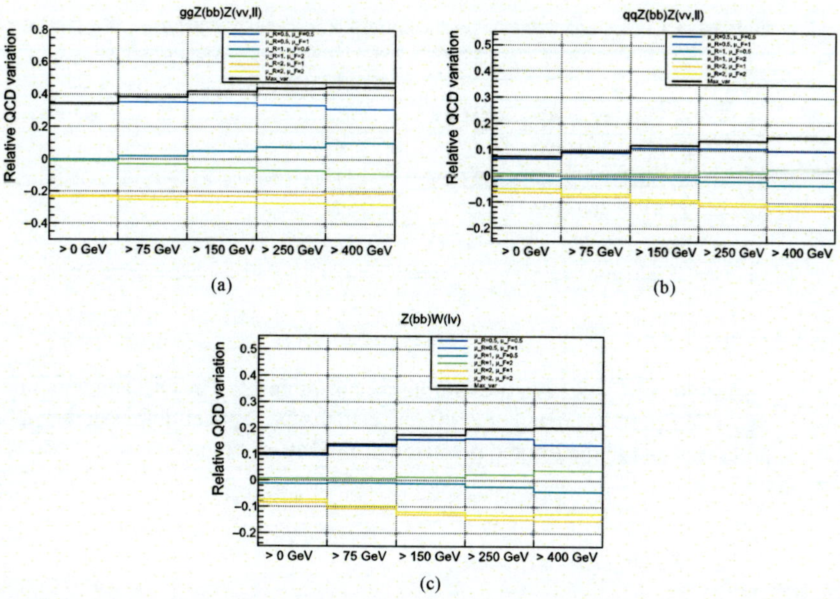

Fig. 9.6 Relative QCD scale variations in inclusive p_{T}^{V} bins for **a** $gg \to ZZ$, **b** $qq \to ZZ$ and **c** $qq \to WZ$ production modes. The *black line* represents the absolute value of the maximum variation and it coincides with one of the six variations

The QCD scale uncertainties are computed by varying the renormalisation scale μ_R and factorisation scale μ_F independently between 1/2 and 2 times their original values μ_R^{norm} and μ_F^{norm}. The six scale variations considered are:

$$\left[\frac{\mu_R}{\mu_R^{norm}}, \frac{\mu_F}{\mu_F^{norm}}\right] : [0.5, 1][1, 0.5][2, 1][1, 2][0.5, 0.5][2, 2] \tag{9.1}$$

The computed variations are relative variations with respect to the nominal values $\left[\frac{\mu_R}{\mu_R^{norm}}, \frac{\mu_F}{\mu_F^{norm}}\right] = [1, 1]$ and they are evaluated in each inclusive p_{T}^{V} region ($p_{\mathrm{T}}^{V} > 0$ GeV, $p_{\mathrm{T}}^{V} > 75$ GeV, $p_{\mathrm{T}}^{V} > 150$ GeV, $p_{\mathrm{T}}^{V} > 250$ GeV, $p_{\mathrm{T}}^{V} > 400$ GeV) for $gg \to ZZ$, $qq \to ZZ$ and $qq \to WZ$ production modes. The absolute value of the maximum of the six variations represents the relative QCD uncertainties in each inclusive p_{T}^{V} bin. Figure 9.6 shows the relative QCD variations of each process together with the absolute value of the maximum variation which coincides with one of the six variations. The absolute value of the maximum variation is of the order of 40–50% for the $gg \to ZZ$ process, and 10–20% for the $qq \to ZZ$ and $qq \to WZ$ processes. Using the Stewart-Tackmann method the absolute values of the maximum variation are calculated in the exclusive p_{T}^{V} bins and they represent the QCD scale variation. The QCD scale variations on the measured STXS bins for the $gg \to ZZ, qq \to ZZ$

Table 9.2 QCD scale uncertainties on the $gg \to ZZ, qq \to ZZ$ and $qq \to WZ$ production modes in the STXS bins where the cross-sections measurements have been performed

Process	δ_{QCD} [%]
$gg \to ZZ$, 250 GeV $\leq p_T^V <$ 400 GeV	50
$gg \to ZZ$, $p_T^V \geq$ 400 GeV	47
$qq \to ZZ$, 250 GeV $\leq p_T^V <$ 400 GeV	17
$qq \to ZZ$, $p_T^V \geq$ 400 GeV	15
$qq \to WZ$, 250 GeV $\leq p_T^V <$ 400 GeV	25
$qq \to WZ$, $p_T^V \geq$ 400 GeV	20

and $qq \to WZ$ processes are summarised in Table 9.2. The QCD uncertainties on the $gg \to ZZ$ process are larger with respect to the QCD uncertainties of the qq-initiated processes due to the limited precision of the MC sample.

9.3.2 PDF Uncertainties

The parton distribution function (PDF) uncertainties have been evaluated using the variations of the NNPDF3.0 set [3]. This PDF set includes a nominal PDF and 100 PDF variations. The PDF uncertainties are computed in each STXS bin for $gg \to ZZ, qq \to ZZ$ and $qq \to WZ$ production modes using the sample standard deviation:

$$\delta_j = \sqrt{\frac{1}{100}\sum_{i=1}^{100}(y_{\text{PDF}i} - y_{\text{PDF0}})^2} \tag{9.2}$$

where y_{PDF0} and $y_{\text{PDF}i}$ are the cross-sections in each STXS bin for the nominal PDF (PDF*0*) and the alternative PDF (PDF*i*). The variation of each alternative PDF is less than 1% in each STXS bin and the overall uncertainty is of the order of 3–4% for the $gg \to ZZ$ process and 1–2% for the qq-initiated processes. Table 9.3 summarises the PDF predictions for the three processes in the p_T^V regions where the cross-section measurements have been performed.

9.3.3 α_S Uncertainties

The α_S uncertainty have been computed using the variations of the NNPDF3.0 set. In addition to the 100 PDF variations with the same nominal value of $\alpha_S(m_Z^2) = 0.118$, the set contains also 2 PDF sets obtained for two alternative values of $\alpha_S(m_Z^2)$: $\alpha_S(m_Z^2) = 0.117$ and $\alpha_S(m_Z^2) = 0.119$. The α_S uncertainty is computed in each STXS bin for the three production modes ($gg \to ZZ, qq \to ZZ, qq \to WZ$) start-

Table 9.3 PDF uncertainties on the $gg \to ZZ$, $qq \to ZZ$ and $qq \to WZ$ production modes in the STXS bins where the cross-sections measurements have been performed

Process	δ_{PDF} [%]
$gg \to ZZ$, 250 GeV $\leq p_{\mathrm{T}}^{V} <$ 400 GeV	2.7
$gg \to ZZ$, $p_{\mathrm{T}}^{V} \geq$ 400 GeV	4.1
$qq \to ZZ$, 250 GeV $\leq p_{\mathrm{T}}^{V} <$ 400 GeV	1.4
$qq \to ZZ$, $p_{\mathrm{T}}^{V} \geq$ 400 GeV	1.8
$qq \to WZ$, 250 GeV $\leq p_{\mathrm{T}}^{V} <$ 400 GeV	1.1
$qq \to WZ$, $p_{\mathrm{T}}^{V} \geq$ 400 GeV	1.8

ing from the relative variations. The relative variations $\mathrm{Var}_{\alpha_S=0.117}$ and $\mathrm{Var}_{\alpha_S=0.119}$ are evaluated using the samples with $\alpha_S(m_Z^2) = 0.117$ and $\alpha_S(m_Z^2) = 0.119$ with respect to the nominal sample. The relative α_S uncertainty is the average of the absolute value of the relative variations $\mathrm{Var}_{\alpha_S=0.117}$ and $\mathrm{Var}_{\alpha_S=0.119}$:

$$\mathrm{Var}_{\alpha_S} = \frac{|\mathrm{Var}_{\alpha_S=0.119}| + |\mathrm{Var}_{\alpha_S=0.117}|}{2} \tag{9.3}$$

Figure 9.7 shows the relative α_S variations and the average of the variations, which is the relative α_S uncertainty, for the three production modes in the STXS bins. The relative α_S uncertainty is of the order of 1–2%. Table 9.4 summarises the relative α_S uncertainties in the STXS regions where the measurements have been performed.

9.3.4 Summary

Thanks to the procedure previously described, the theoretical uncertainties on the theoretical cross-section predictions have been evaluated for the ZZ and WZ processes. Beyond the scope of this thesis, a similar procedure will be applied to evaluate the residual uncertainties necessary to improve the STXS measurements. The total theoretical uncertainty is estimated from the sum in quadrature of the QCD, PFD and α_S uncertainties. The total uncertainties have been evaluated for the $gg \to ZZ$, $qq \to ZZ$ and $qq \to WZ$ processes and then used into each region of the *main splitting*. Table 9.5 shows the total cross-section predictions in the bins of the *main splitting* together with the full set of uncertainties.

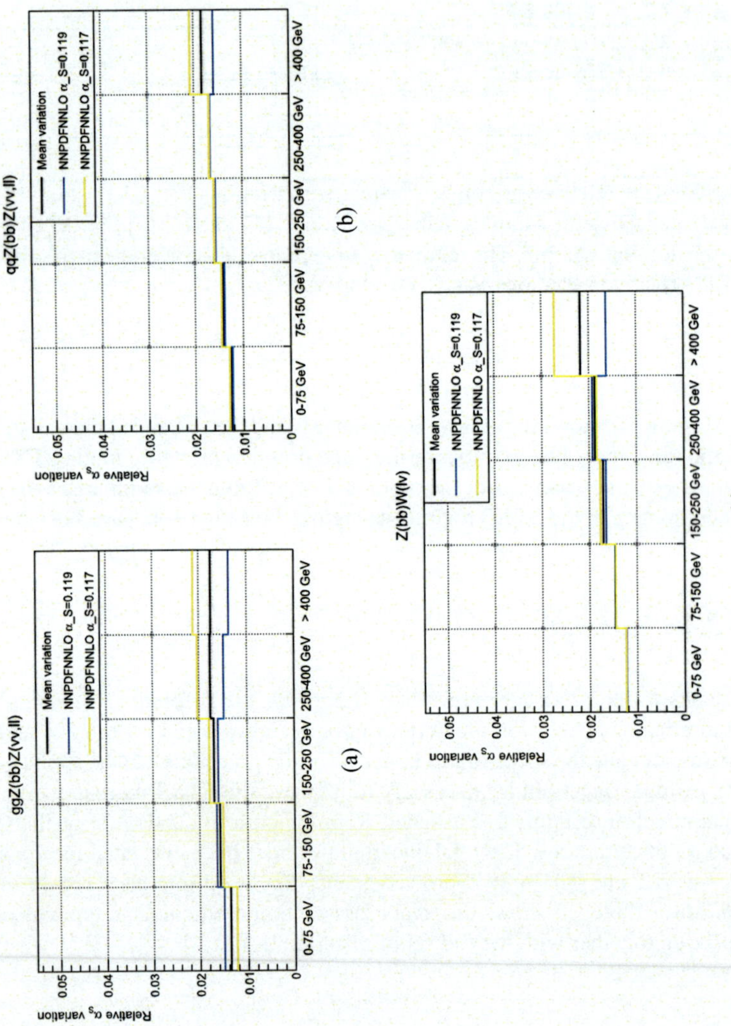

Fig. 9.7 Absolute value of the relative α_S variations in STXS bins for **a** $gg \rightarrow ZZ$, **b** $qq \rightarrow ZZ$ and **c** $qq \rightarrow WZ$ production modes. The *black line* represents the relative α_S uncertainty

Table 9.4 α uncertainties on the $gg \rightarrow ZZ$, $qq \rightarrow ZZ$ and $qq \rightarrow WZ$ production modes in the STXS bins where the cross-sections measurements have been performed

Process	δ_{α_S} [%]
$gg \rightarrow ZZ$, 250 GeV $\leq p_T^V <$ 400 GeV	1.8
$gg \rightarrow ZZ$, $p_T^V \geq$ 400 GeV	1.7
$qq \rightarrow ZZ$, 250 GeV $\leq p_T^V <$ 400 GeV	1.7
$qq \rightarrow ZZ$, $p_T^V \geq$ 400 GeV	1.8
$qq \rightarrow WZ$, 250 GeV $\leq p_T^V <$ 400 GeV	1.9
$qq \rightarrow WZ$, $p_T^V \geq$ 400 GeV	2.2

Table 9.5 Cross-section predictions and theoretical uncertainties for the VZ process at $\sqrt{s}$ = 13 TeV in each region of the *main splitting*

STXS bin	σ [fb]	δ_{tot}^{theo} [fb]	δ_{QCD} [fb]	δ_{PDF} [fb]	δ_{α_S} [fb]
$gg \rightarrow ZZ$, 250 GeV$\leq p_T^V <$ 400 GeV	1.2	0.7	0.6	0.3	0.02
$gg \rightarrow ZZ$, $p_T^V \geq$ 400 GeV	0.2	0.1	0.1	<0.01	<0.01
$qq \rightarrow ZZ$, 250 GeV$\leq p_T^V <$ 400 GeV	14.2	2.4	2.4	0.2	0.2
$qq \rightarrow ZZ$, $p_T^V \geq$ 400 GeV	2.9	0.4	0.4	0.1	0.1
$qq \rightarrow WZ$, 250 GeV$\leq p_T^V <$ 400 GeV	32.7	8.2	8.2	0.4	0.6
$qq \rightarrow WZ$, $p_T^V \geq$ 400 GeV	7.5	1.5	1.5	0.1	0.2

9.4 Results of the Multi-PoIs Fit

The results of the (4+4) PoIs fit are obtained from a binned maximum profile likelihood fit to the data of the m_J distribution. Considering the data collected during the full Run 2 at a center-of-mass energy of $\sqrt{s} = 13$ TeV, corresponding to an integrated luminosity of 139 fb^{-1}, the significances of the VH and VZ processes in each bin are summarised in Table 9.6. The significances of the VH processes are compatible with the ones obtained from the (4+1) PoIs fit (see Table 8.5). The observed VZ significances vary between 1.5 and 2.5σ. Figure 9.8 shows the correlations of the eight parameters of interest which are of the order of 10–20%. The correlations between the VH bins are the same obtained from the (4+1) PoIs fit.

Figure 9.9 shows the comparison of the measured VH and VZ cross-sections with respect to the expected SM values. The theoretical uncertainties on the SM predictions of the cross-sections times branching ratios ($\sigma \times BR$) are shown as grey area. The VZ theoretical uncertainties are bigger than the VH theoretical uncertainties. The large VZ uncertainties are due to the limited precision of some VZ MC samples which are simulated at LO in QCD. All the VH and VZ measurements are in agreement with the SM predictions. The WH and WZ bins have almost the same precision while the ZZ bins has a twice better precision with respect to the ZH bins.

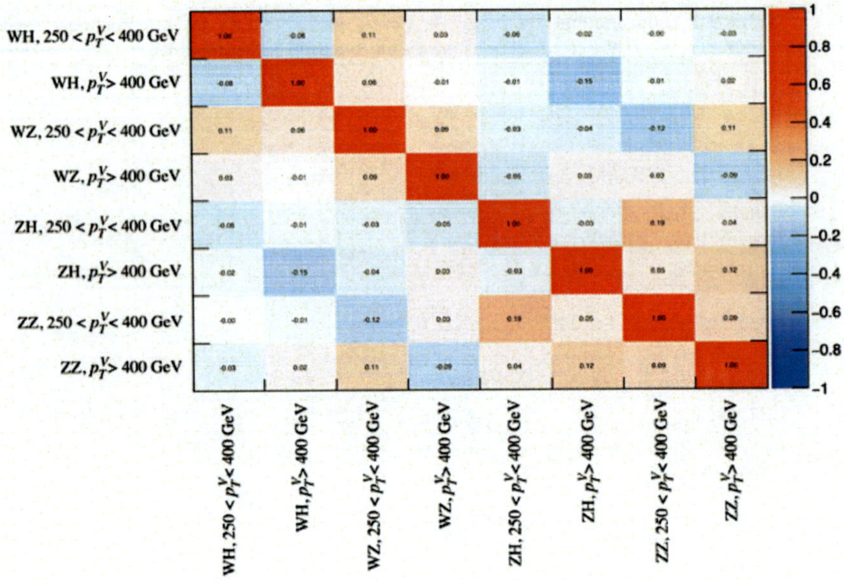

Fig. 9.8 Observed correlations among the cross-sections in the different regions. The colour indicates the size of the correlation

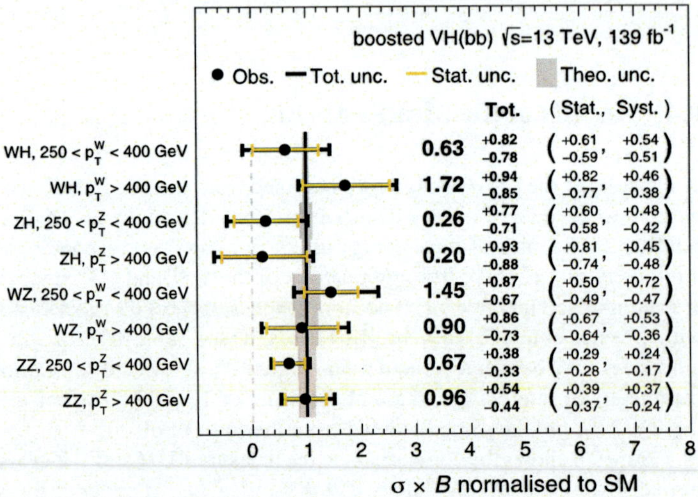

Fig. 9.9 Measured VH and VZ cross-sections times branching ratios divided by the SM predictions. The *grey area* shows the theoretical uncertainty on the SM prediction

Table 9.6 Expected and observed significance for the measured cross-sections in the eight bins

Measurement	Expected sig.	Observed sig.
WH [250, 400[GeV	1.26	0.80
WH [400, ∞[GeV	1.28	2.07
ZH [250, 400[GeV	1.35	0.36
ZH [400, ∞[GeV	1.12	0.22
WZ [250, 400[GeV	2.03	2.44
WZ [400, ∞[GeV	1.34	1.25
ZZ [250, 400[GeV	2.85	2.03
ZZ [400, ∞[GeV	2.39	2.32

Table 9.7 Measured and predicted VZ cross-sections times BRs with corresponding uncertainties in the four STXS bins of the *main splitting* at $\sqrt{s}$ = 13 TeV

Measurement region	SM prediction [fb]	Result [fb]	Stat. unc. [fb]	Syst. unc. [fb]
$Z(\to b\bar{b})Z(\to \ell\ell, \nu\nu)$; $p_T^Z \in$ [250, 400[GeV	15.4 ± 2.5	$10.3^{+5.8}_{-5.1}$	$^{+4.5}_{-4.3}$	$^{+3.7}_{-2.6}$
$Z(\to b\bar{b})Z(\to \ell\ell, \nu\nu)$; $p_T^Z \in$ [400, ∞ GeV	3.1 ± 0.4	$3.0^{+1.7}_{-1.4}$	$^{+1.2}_{-1.1}$	$^{+1.1}_{-0.7}$
$Z(\to b\bar{b})W(\to \ell\nu)$; $p_T^W \in$ [250, 400[GeV	32.7 ± 8.2	48^{+29}_{-22}	$^{+16}_{-16}$	$^{+24}_{-15}$
$Z(\to b\bar{b})W(\to \ell\nu)$; $p_T^W \in$ [400, ∞[GeV	7.5 ± 1.5	$6.8^{+6.4}_{-5.5}$	$^{+5.0}_{-4.8}$	$^{+4.0}_{-2.7}$

The measured VZ cross-sections times $Z \to b\bar{b}$ and $V \to$ leptons branching ratios, together with the SM predictions, are summarized in Table 9.7 and illustrated in Fig. 9.10. The VZ cross-sections are measured with relative uncertainties varying between 60 and 95% with almost the equal contribution of the statistical and systematic uncertainties. The analysis has large VZ systematic uncertainties which are over-conservative and highly ranked in the VZ bins. The next natural step of the analysis is to improve the VZ uncertainty computation considering only residual uncertainties while performing the cross-section measurements. The VZ contribution on the systematic uncertainties will likely go down.

In this new measurement, the NPs of all the systematics have been fully scrutinized. The obtained NPs have been compared to the ones of the (4+1) PoIs fit. The pulls and the constraints of all the NPs do not change. In some cases some NPs show up in the pull plot because they are not pruned any more. In these particular cases the NP is neither pulled nor constrained. Furthermore the normalization factors of the major backgrounds (V+jets and $t\bar{t}$) are the same.

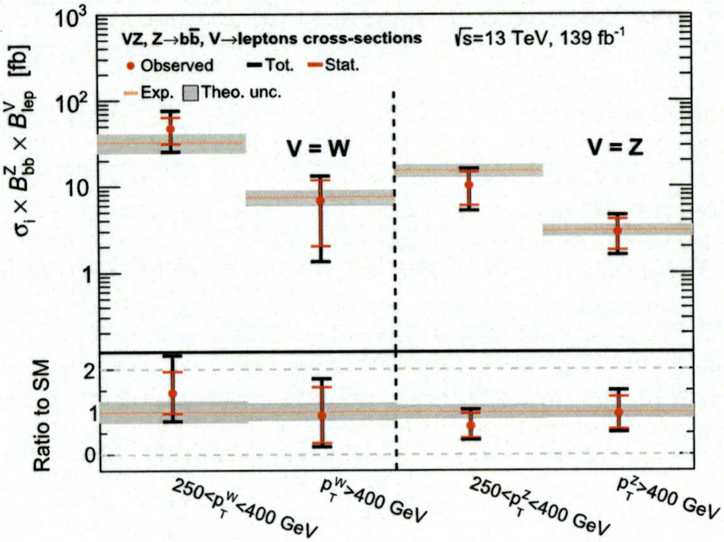

Fig. 9.10 Measured VZ cross-sections times branching ratios in the four STXS bins at $\sqrt{s} = 13$ TeV

9.4.1 What's Next?

As shown in this chapter, it is possible to performed a VZ cross-section measurement starting from the $VH(b\bar{b})$ *boosted* analysis. Nevertheless a more reliable study of the VZ systematic uncertainties is needed since now they are over-conservative. The new VZ uncertainty scheme will be based on the same methods applied for the theoretical VZ uncertainties and it will give a more reliable estimation of the systematic uncertainties specifically needed for the cross-section measurements.

The next natural step is the EFT interpretation of the results. The VZ cross-section can be parametrized as a function of Wilson coefficients while performing the (4+4) PoIs fit. Preliminary results shows that parametrizing the VH and VZ cross-sections with the Wilson coefficients, there is an improvements of 5–10% of the expected Wilson coefficient limit. The preliminary tests have been done performing a one-dimensional fit and considering the linear and quadratic parametrisation of the cross-section. The results looks promising and can help to improve limits for some operators.

References

1. Chatrchyan S et al (2014) Measurement of WZ and ZZ production in pp collisions at vs = 8 TeV in final states with b-tagged jets. Eur Phys J C 74:2973. https://doi.org/10.1140/epjc/s10052-014-2973-5 arXiv: 1403.3047 [hep-ex]
2. Stewart IW, Tackmann FJ (2012) Theory uncertainties for Higgs and other searches using jet bins. Phys Rev D 85:034011. https://doi.org/10.1103/PhysRevD.85.034011 arXiv: 1107.2117 [hep-ph]
3. Ball RD et al (2015) Parton distributions for the LHC Run II. JHEP 04:040. https://doi.org/10.1007/JHEP04(2015)040 arXiv: 1410.8849 [hep-ph]

Chapter 10
Conclusions

The discovery of the Higgs boson represents an outstanding achievement of particle physics. In the last decade, precise measurements have probed the predictions of the Standard Model and its possible extensions. Using the data collected by LHC, all the measurements are compatible with the Standard Model predictions and no hint of new physics has been found.

This thesis is focused on the search for the Standard Model $H \rightarrow b\bar{b}$ decay which is an important Higgs boson decay channel to establish the Yukawa nature of the Higgs boson to fermions couplings. The observation of this decay at LHC has been obtained only after seven years from the Higgs boson discovery because this channel is affected by large backgrounds arising from multi-jet production in the dominant gluon-gluon fusion production mode. The most sensitive production modes for detecting the $H \rightarrow b\bar{b}$ decay are the associated production of the Higgs boson with a Z or a W boson. The suppression of the multi-jet background is achieved by the leptonic decay of the vector boson.

The search of the $VH(b\bar{b})$ production has been developed by two complementary analysis strategies, the $VH(b\bar{b})$ *resolved* analysis and the $VH(b\bar{b})$ *boosted* analysis. In the $VH(b\bar{b})$ *resolved* analysis the Higgs boson decay products are reconstructed as two separate b-jets, while in the *boosted* analysis the Higgs boson decay products tends to merge in a large-R since the analysis is designed for the high energy phase space. Using the full Run 2 dataset, the $VH(b\bar{b})$ *resolved* analysis has observed the Higgs boson production in association with a Z boson and has measured a strong evidence of the Higgs boson associated production with a W boson. Furthermore, using the Simplified Template Cross-Section framework the ZH and WH cross-sections have been measured as a function of the gauge boson transverse momentum in kinematic fiducial volumes.

The analysis presented in this thesis is the $VH(b\bar{b})$ *boosted* analysis, a new version of the analysis designed to probe the Higgs boson when it is produced with a very large transverse momenta. To enhance the sensitivity, the Higgs boson is reconstructed using the large-R jet technique. The analysis results are based on the Run 2 dataset

G. Di Gregorio, *A Beauty-ful Boson*, Springer Theses,
https://doi.org/10.1007/978-3-031-20013-7_10

of p-p collision data collected at $\sqrt{s} = 13$ TeV by the ATLAS detector, corresponding to an integrated luminosity of 139 fb^{-1}. For a Higgs boson mass $m_H = 125$ GeV, an excess over the expected Standard Model background is observed with a significance of 2.1 standard deviations (2.7 standard deviations expected). The result is the most precise measurement currently available in the high transverse momentum regime for this process. Given that the analysis is limited by the statistical uncertainties, the next round of data collection (Run 3) will be important to improve the analysis sensitivity.

The $VH(b\bar{b})$ *boosted* analysis performs a simultaneous fit to test the analysis on the VZ irreducible background which has a similar topology to the VH signal. For the VZ process, a significance of 5.4 standard deviations is observed, compared to an expectation of 5.7 standard deviations. The VZ result provides a direct observation of the $VZ(b\bar{b})$ production mode.

The encouraging results favour performing cross-section measurements in the Simplified Template Cross-Section framework. The cross-sections are measured separately for the ZH and WH productions in the two p_{T}^{V} regions, 250 GeV $\leq p_{\mathrm{T}}^{V} < 400$ GeV and $p_{\mathrm{T}}^{V} \geq$ GeV. All the measurements are in agreement with the SM expectations. Searches for possible deviations from Beyond the Standard Model physics are performed using an effective field theory approach. Limits are set on the coefficients of the effective Lagrangian operators which affect the VH production and $H \to b\bar{b}$ decay. Extending the STXS scheme with a cut on $p_{\mathrm{T}}^{V} = 400$ GeV, a 70% improvement on the confidence intervals is extracted from one-dimensional and two-dimensional fits.

An alternative approach, based on the $VH(b\bar{b})$ *boosted* analysis, is presented. This analysis performs cross-section measurements of the diboson ZZ and WZ processes. For the first time the $ZZ(b\bar{b})$ and $WZ(b\bar{b})$ cross-sections are measured at $\sqrt{s} = 13$ TeV, with relative uncertainties varying between 60% and 95%. The results are consistent with the Standard Model predictions. Further studies on the modelling of the VZ process are the key ingredients to improve the analysis sensitivity. The obtained results can be reinterpreted using an effective field theory approach.

The final step of the $VH(b\bar{b})$ *boosted* analysis will be the combination with the $VH(b\bar{b})$ *resolved* analysis. The phase space of the two analyses significantly overlaps and the combination is not straightforward. About 60% of the VH signal events are reconstructed by both analyses. Several possibilities are available to implement such a combination and the choice will depend on the expected performance. More detailed studies on the events reconstructed by both analyses are needed to decide the combination strategy.

Another important future developments is the investigation using machine learning techniques as an alternative to the cut-based procedure adopted in the $VH(b\bar{b})$ *boosted* analysis. A possibility will be to introduce a boosted decision tree as it is done at the moment in the *resolved* analysis. Another chance is to exploit new machine learning techniques to use in the combination to enhance the analysis sensitivity.

Last but not least, an important aspect to explore in the combined analysis is the possibility to extend the STXS scheme. Combining the two analyses there is an increase of statistics and the STXS bins can be split according to the number extra jets in the events. The new STXS scheme can be used to investigate on possible effects from Beyond the Standard Model physics.

The current picture of the Standard Model is far from being complete. Precision measurements of the VH production and $H \to b\bar{b}$ decay provide an effective way to search for Beyond the Standard Model physics.

Appendix A
EFT Eigenvector Results in the $VH(b\bar{b})$ *Resolved* Analysis

Another method to extract information on possible BSM physics, tested by the $VH(b\bar{b})$ *resolved* analysis, is based on the use of eigenvectors. The eigenvectors contain information about the sensitivity of the analysis to various deformation of the Standard Model. Since there are five STXS bins in the *resolved* analysis, five eigenvectors are defined. The definition of the eigenvectors is shown in Table A.1 where they are ordered in terms of experimental sensitivity, the highest for c_{E0} and the lowest for c_{E4}. The leading eigenvector c_{E0} consists almost exclusively of c_{Hq3} which is mostly constrained by the 1D fit. The second eigenvectors c_{E1} is dominated by c_{Hu} but has sizeable contributions from c_{Hd} and c_{Hq1}. The eigenvector c_{E2} demonstrates sensitivity to a combination of the branching ratio independent parameter $\mathcal{I}_{BR}$ and c_{HW}, while c_{E3} has a limited sensitivity to a combination of c_{HWB} and c_{Hq1}. The analysis has a negligible sensitivity to the fifth eigenvector c_{E4}. The result of simultaneous fit to the five leading eigenvectors is shown in Table A.2. Using the fit results the impact of the four leading eigenvectors on the cross-section measurements in each STXS bins has been studied (Fig. A.1). The c_{E0} and c_{E1} coefficients induce bigger deviations from the SM in the higher p_{T}^{V} STXS bins, while c_{E2} and c_{E3} have opposite effects.

Table A.1 Linear combination of the Wilson coefficients used to define the eigenvectors. All the modifications that alter the branching ratio are observed into an additional independent term ($\mathcal{I}_{BR}$) which linearly change the branching ratio. All the contributions with a coefficient below 0.2 are omitted [1]

Eigenvector name	Eigenvector definition
c_{E0}	$0.98 \cdot c_{Hq3}$
c_{E1}	$0.85 \cdot c_{Hu} - 0.39 \cdot c_{Hq1} - 0.27 \cdot c_{Hd}$
c_{E2}	$0.70 \cdot \mathcal{I}_{BR} + 0.62 \cdot c_{HW}$
c_{E3}	$0.74 \cdot c_{HWB} + 0.53 \cdot c_{Hq1} - 0.32 \cdot c_{HW}$
c_{E4}	$0.65 \cdot c_{HW} - 0.60 \cdot \mathcal{I}_{BR} + 0.35 \cdot c_{Hq1}$

G. Di Gregorio, *A Beauty-ful Boson*, Springer Theses,
https://doi.org/10.1007/978-3-031-20013-7

Table A.2 Expected and observed best-fit values and associated 1σ uncertainties from a simultaneous fit of the five leading eigenvectors

Eigenvector	Expected	Observed
c_{E0}	$0.0^{+0.024}_{-0.022}$	$0.005^{+0.023}_{-0.022}$
c_{E1}	$0.0^{+0.164}_{-0.162}$	$0.058^{+0.163}_{-0.161}$
c_{E2}	$0.0^{+0.346}_{-0.336}$	$-0.10^{+0.35}_{-0.34}$
c_{E3}	$0.0^{+2.07}_{-2.06}$	$0.696^{+2.08}_{-2.05}$
c_{E4}	$0.0^{+11.3}_{-11.5}$	$2.3^{+12.5}_{-12.6}$

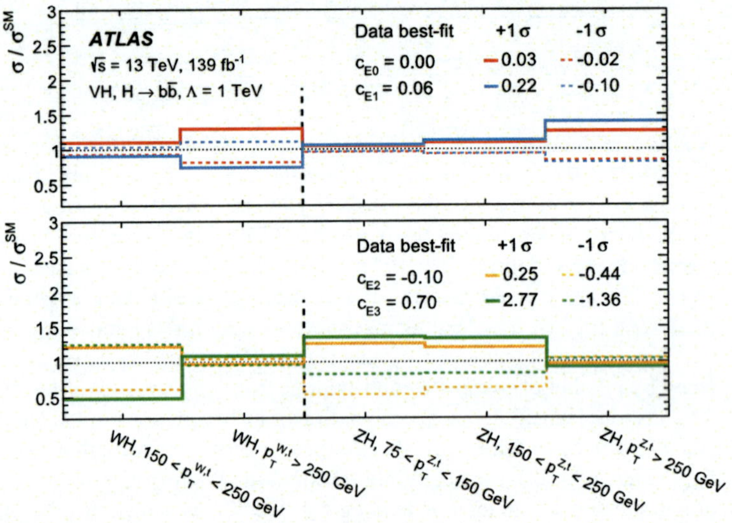

Fig. A.1 Impact of the leading four eigenvectors on the STXS cross-section measurements. The change of the cross-section is indicated at $+1\sigma$ (*solid line*) and -1σ (*dashed line*) limits of the corresponding coefficient. The value of the coefficients are extracted from a simultaneous fit of all five eigenvectors [1]

Reference

1. Aad et al G (2020) Measurements of W H and Z H production in the $H \to b\bar{b}$ decay channel in pp collisions at 13 TeV with the ATLAS detector. arXiv: 2007.02873 [hep-ex]

Appendix B
The Tile Calorimeter Design, Calibration Methods and the Photomultiplier Stability

This appendix briefly describes the Tile Calorimeter structure, together with the calibration methods used to monitor the stability and the performance of each part of the read-out chain. The second part of the appendix is focused on the response stability of the photomultiplier reading out the Tile Calorimeter.

B.1 The Tile Calorimeter Layout

The Tile Calorimeter is the central section ($|\eta| < 1.7$) of the hadronic calorimeter made up of steel plates, as absorbers, and plastic scintillating tiles, as active materials. The calorimeter is subdivided in three regions along the z-axis: a central section called Long Barrel, and two external sections called Extended Barrel (EB). The LB covers the region $|\eta| < 1$, while the EB covers two regions of $0.8 < |\eta| < 1.7$. Full azimuthal coverage around the beam axis is achieved with 64 wedge-shaped modules, each covering $\Delta\phi = 0.1$ radians.

As shown in Fig. B.1, the tiles are perpendicular to the beam axis and the light produced by the scintillating tiles at particle crossing is collected by wave-length shifting fibres. The fibres deliver the light to the two photomultipliers (PMTs)[1] located in the outer radius iron structure that also houses the front-end electronics. The PMT outputs are amplified, shaped and finally digitized by ADCs, and then stored in the front-end pipeline memory.

TileCal is subdivided in three separate longitudinal sampling layers, called respectively A, BC (just B in the EB) and D. TileCal cells, shown in Fig. B.2, are defined by grouping the wave-leght shifting fibres from individual tiles on the corresponding PMT. The granularity of each cell is $\Delta\eta \times \Delta\phi = 0.1 \times 0.1$ for the A and

[1] The photomultiplier is a device which converts light pulses into electron current pulses through a photo-electron emission and charge amplification.

G. Di Gregorio, *A Beauty-ful Boson*, Springer Theses,
https://doi.org/10.1007/978-3-031-20013-7

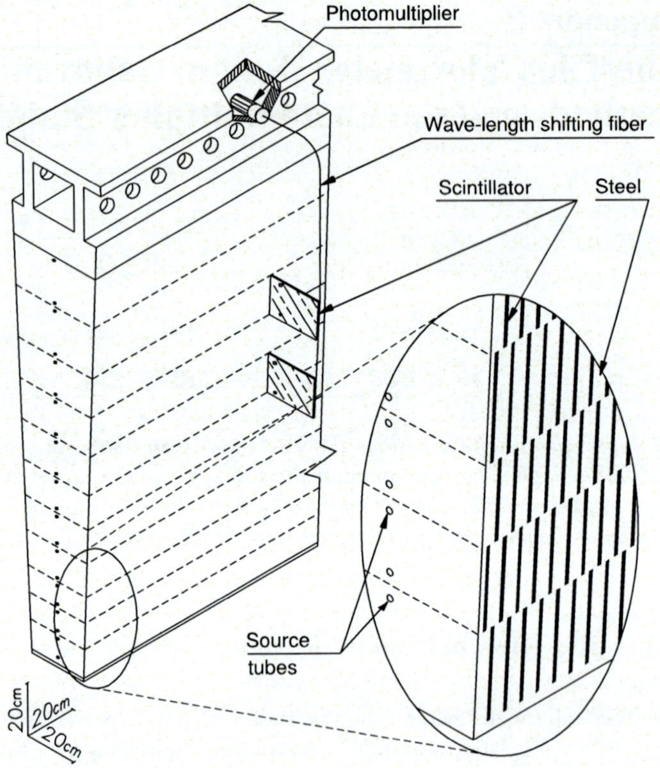

Fig. B.1 Illustration of the mechanical assembly and optical read-out of a single TileCal modules [1]

BC (or B) innermost layers, while the cells of the D layer have a granularity of $\Delta\eta \times \Delta\phi = 0.2 \times 0.1$. Between the LB and the EB there are the E cells which are only composed of scintillator and exceptionally read out by only one PMT.

B.2 Calibration Methods

To monitor the TileCal response stability a continuous multi-stage calibration procedure is used. Three calibration systems are used to monitor the stability of each channel of the read-out chain and to provide the correct calibration. The reconstructed energy of each channel of TileCal, E [GeV], is expressed in term of the ADC raw response, A[ADC], as it follows:

$$E[\text{GeV}] = A[\text{ADC}] \cdot C_{\text{ADC}\rightarrow\text{pC}} \cdot C_{\text{pC}\rightarrow\text{GeV}} \cdot C_{\text{Cesium}} \cdot C_{\text{Laser}} \quad \text{(B.1)}$$

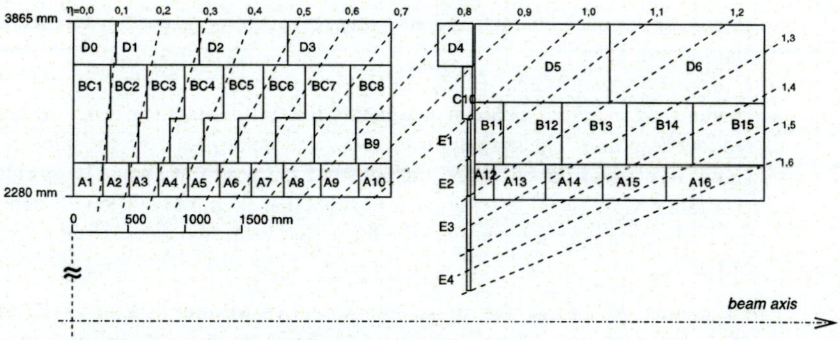

Fig. B.2 The layout of the TileCal cells, denoted by a letter (A to E) plus an integer number. The A-layer is closest to the beam-line [1]

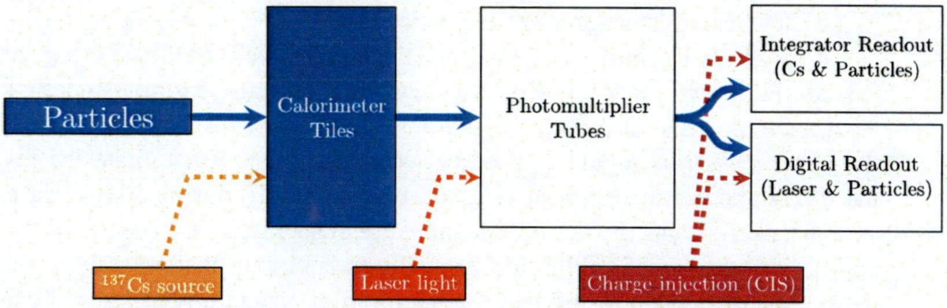

Fig. B.3 The signal paths for each of the three calibration systems used by TileCal. The physics signal is denoted by the *solid line*, while the path taken by each calibration system is shown with *dashed lines* [1]

The conversion factor from pC to GeV, $C_{\mathrm{pC}\rightarrow\mathrm{GeV}}$, was fixed during the test beam campaigns using particles with known energy that cross the calorimeter. The remaining calibration constants are provided by the three calibration systems, as shown in Fig. B.3:

- *Cesium calibration system*: this calibration system is based on a movable radioactive sources which is moving through the calorimeter using a hydraulic system. The response of all TileCal cells is equalized with a Cesium system by monitoring the cell and the PMT response to the source emission. This calibration system provides the C_{Cesium} constant.
- *Laser calibration system*: this calibration system consists of a laser source that sends light pulses to all the PMTs through an optical system. The laser light pulses are similar to those produced by ionizing particles. The systems is used to monitor and to correct the PMT response variations between two Cesium scans. Laser calibrations are performed twice per week when particles are not colliding in the LHC. The Laser system provides the laser calibration constant C_{Laser} which is

calculated for each channel and with respect to a reference run taken just after the last Cesium scan.

- *Charge Injection system* (CIS): this system simulates a physical signal in the channels by injecting a known charge into the ADCs and measuring the electronic readout response. The system provides a relationship between the injected analog signal in pC and the electronic response of the read-out channels. A calibration constant $C_{\mathrm{ADC}\rightarrow\mathrm{pC}}$ is determined for each TileCal channel. The CIS calibrations are performed during each machine filling between two consequent physics runs of LHC.

Additionally, the use of the signal integrators in Minimum Bias (MB) events is another diagnostic tool developed to cross-check the previous calibration systems. The minimum bias integration is based on measuring the detector activity induced by *p*-*p* collisions at small transferred momentum. The rate of the MB events is proportional to the instantaneous luminosity and these particular events can be used to monitor the calorimeter response variation.

The absolute calibration of the TileCal energy scale is obtained using the Cesium system. However, since the Cesium scans need a pause in the *p*-*p* collisions of at least six hours, this calibration procedure cannot be performed very often. Therefore, the Laser system is regularly used between two scans to calibrate the response of the PMTs and of the electronic readout. Unfortunately, during 2016 a water leakage was discovered and the Cesium scans were suspended. In absence of the Cesium calibration, the Laser calibration procedure acquires a leading role in the TileCal calibration. A description of the Laser calibration system is reported in the following.

To maintain the calorimeter high performance, a continuous calibration of each cell is required. The stability of the calorimeter response, within 1%, is important for the jet and missing transverse momentum reconstruction. TileCal reconstructs up to 40% of the energy of the jets with $p_{\mathrm{T}} > 3$ TeV. The mean fraction of the contained jet energy increases with increasing jet p_{T} from approximately 20–40% between 100 and 3 TeV. To obtain a precise and stable measurement of the energy deposited in the calorimeter, it is mandatory to precisely monitor any variation of the TileCal response. If there is a response variation, a correction is applied to restore the correct value and maintain the stability.

B.2.1 Laser Calibration System

The Laser system is used to monitor and to calibrate the PMT response. The system was upgraded during the shutdown before the beginning of the Run 2 to improve the system calibration reliability and to maintain the signal calibration at sub-percent level accuracy.

A sketch of the Laser II system is shown in Fig. B.4. The path of the laser light starts in the optics box. The light emitted by laser head is split into two beams by a beam splitter. A few percent of the reflected light is sent to three photodiodes, while

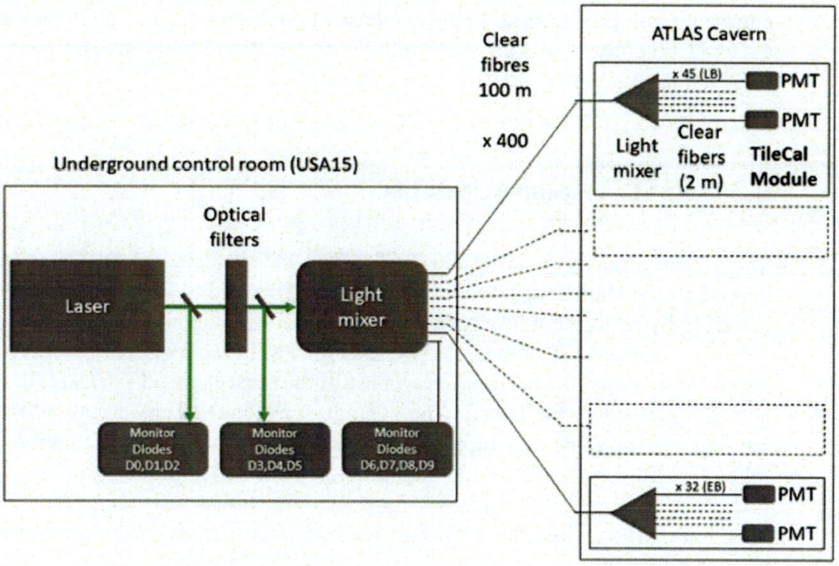

Fig. B.4 Sketch of the Laser II System

the main beam is transmitted into a filter wheel. The filter wheel houses different optical density filters which are used to vary the intensity of the transmitted light. After the filter wheel another beam splitter is placed. A fraction of the light is sent to other three photodiodes, while the remaining fraction enters into a light mixer in order to mix light and to expand the beam. The light at the light mixer exit is then collected by a bundle of fibres. Four of these fibres are connected to four photodiodes, while the other fibres, 100 m long, transmit the light to 384 light mixers located in each TileCal module. The light from each light mixer is then sent to all the PMTs reading out a TileCal module.

In the Laser calibration system the photodiodes are used to probe the laser beam stability at different steps of the light transmissions. The stability of the photodiodes is monitored with a LED whose stability is controlled by a reference photodiode. The reference photodiode is equiped with a radioactive source that allows to check independently its own stability.

The laser pulses have a time profile similar to those of light signals generated by particle crossing in the calorimeter cells. However, the optical path is different in the two cases and the light transmission stability of the optical path for the laser beam has to be controlled. A specific algorithm measures the drifts due to the transmission system only. A detailed description of the algorithm can be found in Ref. [2]. The algorithm is used to evaluate the Laser calibration constant C_{Laser} which is applied to correct for non-uniformities in the response of the PMTs of TileCal. In the following the PMT response variation measured during the Run 2 and the expected response variation during the HL-LHC era are presented.

B.3 PMT Response Stability

TileCal is readout by about 10,000 PMTs. The PMTs selected for TileCal are a special version of the Hamamatsu model R5900. The TileCal PMT version is R7877 characterized by response linearity in a huge interval of incident light intensity and fast rise time. The Laser system is used to study the PMT response evolution. Figure B.5a shows the average PMT response variation of the most exposed TileCal cells (A-cells and E-cells) during the Run 2. The average is evaluated considering all the PMTs reading out the same cell type. The response is normalised to the calibration run taken on July 17th 2015, at the beginning of Run 2. The PMT response evolution is characterized by down-drifts and up-drifts. The down-drifts coincide with the *p-p* collision periods while the response recovery occurs during the technical stops of the LHC machine.

Moreover, the average PMT response can be studied as a function of the integrated anode charge, as shown in Fig. B.5b. The response is normalised to the calibration run taken before the beginning of Run 2 (July 17th 2015). The points corresponds to the average response of the PMTs reading the most exposed cells. The behaviour of the PMT response changes when large currents are integrated. The average PMT response is fitted with a double exponential function $R(Q)$, defined as:

$$R(Q) = p_0 \cdot \exp(p_1 \cdot Q) + p_2 \cdot \exp(p_3 \cdot Q) \tag{B.2}$$

where Q is the integrated anode charge. The double exponential function $R(Q)$ describes fairly well both the initial part of the response evolution and the high value region of the integrated charge. The double exponential function can be used

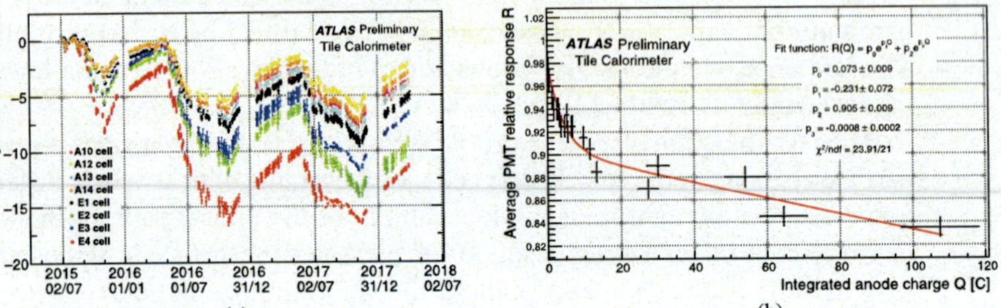

(a) (b)

Fig. B.5 Average PMT response variation of the most exposed TileCal cells as a function of time (**a**) and of the integrated anode charge (**b**) during the Run 2 [1] to make projections about the expected PMT response loss in the HL-LHC. The expected integrated luminosity at the end of the HL-LHC is 4000 fb^{-1}, corresponding to 600 C of integrated anode charge for the PMTs looking to the light output of the most exposed cells. The expected response loss of the PMTs reading the most exposed cells is more than 25%. These PMTs represent 8% of total 10,000 TileCal PMTs and they will be replaced by a newer version but with same geometry. The new PMT version is model R11187 which have been tested in the Pisa-INFN laboratory.

B.3.1 PMT Ageing

A local test bench to evaluate the effect of the charge integration of the PMTs is operating at Pisa-INFN laboratory. The optical system used in the laboratory is similar to the one used in TileCal for the Laser calibration. In the laboratory, the anode charge integration is done with a pulsing green LED, while the PMT response is measured with the laser. The current (R7877) and the new (R11187) PMT model were tested in the laboratory. The tested R7877 PMTs were dismounted from the TileCal detector in February 2017 and were reading different cell types having integrated 1 to 5 C since the beginning of the LHC operating period. Figure B.6 shows the average PMT response as a function of the integrated anode charge for the two PMT models. The blue points represent the average response of 7 PMTs model R7877, while the red triangular points represent the average response of 4 PMTs model R11187. The PMT response is normalised to the first day of observation. The average PMT response of the two PMT groups is fitted with the double exponential function (Eq. B.2). The new PMT model has a smaller down-drift with respect to the old PMT version.

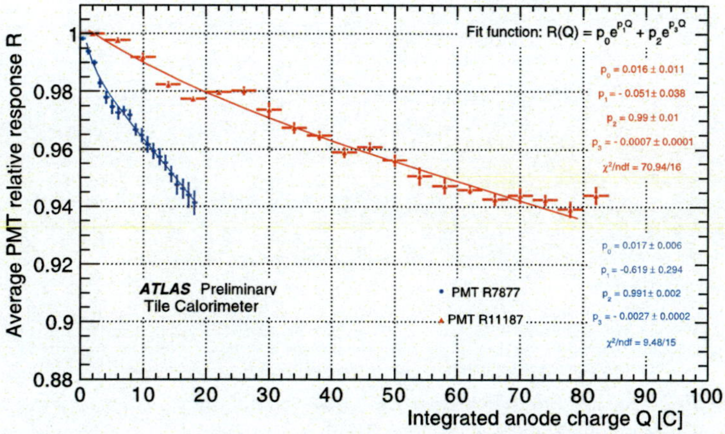

Fig. B.6 Average PMT relative response of the old (*blue*) and new (*red*) PMT models at the test bench [1]

In conclusion, calibration data and test bench data provide consistent results showing that the PMTs are affected by a degradation of their response. The response degradation depends on the amount of integrated anode charge. Moreover the calibration data provide useful information to predict the PMT behaviours during the HL-LHC period. To prevent large response losses of PMTs reading the most exposed cells, 8% of the total TileCal PMTs will be replaced with a new generation PMTs with improved performance.

References

1. ATLAS Collaboration, TileCal Public Plots, url: http://twiki.cern.ch/twiki/bin/view/AtlasPublic/ApprovedPlotsTile
2. G. Di Gregorio, Studies of the response stability for long term photomultiplier operation in the ATLAS hadronic calorimeter and a new method for photomultiplier gain measurements

Appendix C
Efficiency Studies in the Event Selection

In this appendix, all the efficiency studies in the 0-lepton channel are described in more details. The impact of each cut applied in the event selection is shown.

C.1 Efficiency Studies in 0-Lepton Channel

The impact of a single cut applied in the event selection is studied calculating the efficiency defined as the number of events that pass that specific cut to the number of events that pass the previous cut. Figure C.1 shows the efficiency of all the cuts, except for the b-tagging requests, applied in event selection in the 0-lepton channel as a function of the truth p_{T}^{V}. The efficiency of the b-tagging algorithm is already described in Sect. 5.2.2.

Figure C.1a shows the efficiency of the $E_{\mathrm{T}}^{\mathrm{miss}}$ cut evaluated as the ratio of the number of events passing the *trigger* and E_T^{miss} *cut* over the number of events passing the *trigger cut*. The trend of E_T^{miss} *cut* is characterized by a sharp turn on at around truth $p_{\mathrm{T}}^{V} \sim 250$ GeV and then at $p_{\mathrm{T}}^{V} \sim 400$ GeV the E_T^{miss} *cut* reaches a plateau and it is fully efficient.

Figure C.1b shows the efficiency of the *large-R jet cut* calculated as the ratio of the number of events passing the *trigger*, E_T^{miss} and *large-R jet cuts* over the number of events passing only the *trigger* and E_T^{miss} *cuts*. The plots as a minimum at truth $p_{\mathrm{T}}^{V} \sim 250$ GeVwhich is probably connected to the cut on the p_{T} of the large-R jet.

Figure C.1c shows the efficiency of the *anti-QCD cut* evaluated as the ratio of the number of the events passing the *trigger*, E_T^{miss}, *large-R jet* and *anti-QCD cuts* over the number of events passing the *trigger*, E_T^{miss} and *large-R jet cuts*. The *anti-QCD cut* has a flat trend and it discards about 8% of the events.

Figure C.1d shows the efficiency of the *VR track-jets cut* evaluated as the ratio of the number of events passing the *trigger*, E_T^{miss}, *large-R jet*, *anti-QCD* and *VR track-jets cuts* over the number of events passing the *trigger*, E_T^{miss}, *large-R jet* and

G. Di Gregorio, *A Beauty-ful Boson*, Springer Theses,
https://doi.org/10.1007/978-3-031-20013-7

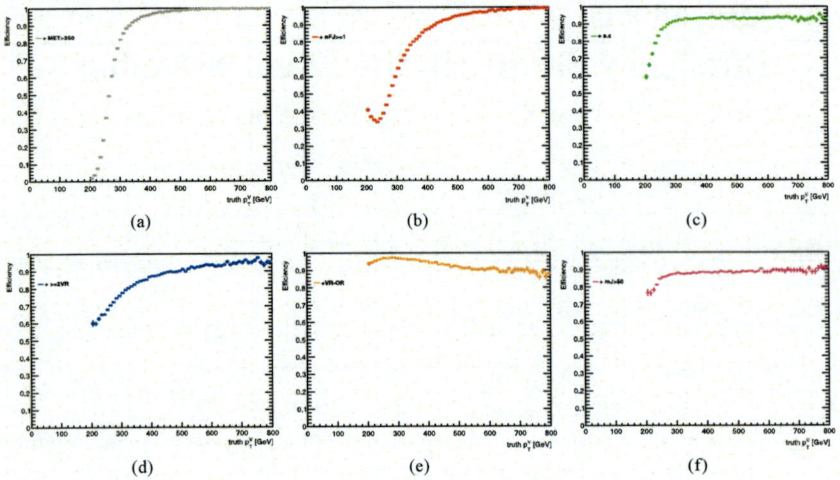

Fig. C.1 Efficiency breakdown in 0-lepton channel. **a** Events that pass the E_T^{miss} *cut* ($E_{\mathrm{T}}^{\mathrm{miss}} > 250$ GeV) divided by the events that pass the *trigger cut*. **b** Events have at least one large-R jet, that pass the E_T^{miss} and *trigger cuts* divided by the events that pass the *trigger* and $E_{\mathrm{T}}^{\mathrm{miss}}$ *cuts*. **c** Events that pass the *trigger*, E_T^{miss}, *large-R jet* and *anti-QCD cuts* divided by the events that pass *trigger*, $E_{\mathrm{T}}^{\mathrm{miss}}$ and *large-R jet cuts*. **d** Events that pass the *trigger*, E_T^{miss}, *large-R jet*, *anti-QCD* and *VR track-jets cuts* divided by the events that pass *trigger*, E_T^{miss}, *large-R jet* and *anti-QCD cuts*. **e** Events that pass the *trigger*, E_T^{miss}, *large-R jet*, *anti-QCD*, *VR track-jets* and *VR overlap removal cuts* divided by the events that pass the *trigger*, E_T^{miss}, *large-R jet*, *anti-QCD* and *VR track-jets cuts*. **f** Events that pass the *trigger*, E_T^{miss}, *large-R jet*, *anti-QCD*, *VR track-jets*, *VR overlap removal* and m_J *cuts* divided by events that pass the *trigger*, E_T^{miss}, *large-R jet*, *anti-QCD*, *VR track-jets* and *VR overlap removal cuts*

anti-QCD cuts. As expected the efficiency of the number of VR track-jets tends to increase in the high energy region from 70 to 90% since at high energy more jets are produced.

Figure C.1e shows the efficiency of the *VR overlap removal cut* calculated as the number of events passing the *trigger*, E_T^{miss}, *large-R jet anti-QCD*, *VR track-jets* and *VR overlap removal cuts* over the number of events passing the *trigger*, E_T^{miss}, *large-R jet anti-QCD* and *VR track-jets cuts*. The efficiency of the VR overlap removal procedure decreases at high energy because the VR track-jets start to merge so more events are discarded.

Figure C.1f shows the efficiency of the m_J *cut* calculated as the number of events passing the *trigger*, E_T^{miss}, *large-R jet*, *anti-QCD*, *VR track-jets*, *VR overlap removal* and m_J *cuts* over the number of events passing the *trigger*, E_T^{miss}, *large-R jet*, *anti-QCD*, *VR track-jets* and *VR overlap removal cuts*. The plot has a flat trend in full range and the efficiency is 90%.

Appendix D
Pull Plots

The pull plots shown in this appendix are obtained from the unconditional fit of the three lepton channels to Asimov dataset (red dots) and to actual dataset (black dots). Figure D.1 shows the pull plots of the NPs associated to the modelling uncertainties of background ($Z+$ jets, $W+$ jets, diboson, QCD) processes and signal process.

Figure D.1a and b show the pull plots for the $Z+$ jets and $W+$ modelling uncertainties. The `norm_Zbb` and `norm_Wbb` NPs are the floating normalisation factors for Z+HF and W+HP events, respectively. The best-fit floating normalization factor for Z+HF (W+HF) events is 1.32 (1.12), as reported in Table 8.2. The fit tends to increase the number of Z+HF (W+HF) events. The `R_HPLP`, `R_Ptv`, `R_SRCR` NPs describe the acceptance uncertainties between the HP and LP SR, between the medium and high p_{T}^{V} region and, between the SR and CR, respectively. The `R_SRCR_CR_Zhf` and `R_SRCR_CR_Whf` NPs are pulled up, while the `R_HPLP_SR_J0_Whf`, `R_Ptv0L_400_Whf` and `R_Ptv1L_400_Whf` NP are pulled down. The pulls of the NP without any constraints can occur because the fit tends to adjust the MC prediction to the observed data. The remaining NPs in the plots describes the shape uncertainties (`ZmJShape`, `WmJShape`), the normalisation uncertainty of the $Z \rightarrow b\bar{b}$ and $W \rightarrow b\bar{b}$ events and minor background processes (i.e. `ZbbNorm_L0`, `ZclNorm`, `ZlNorm`), and the flavour composition uncertainties (i.e. `ZbcZbbRatio`, `ZblZbbRatio`, `ZccZbbRatio`).

Figure D.1c, d and e show the pull plots of the diboson, QCD and signal VH modelling uncertainties. The plots do not show any suspicious behaviour.

Figure D.2 shows the pull plots of the detector-related nuisance parameters obtained from the unconditional fit in all the three lepton channels to the Asimov dataset and actual dataset. Figure D.2a shows the NPs related to the small-R jets uncertainties which are expected to have a small impact because the small-R jets are used only in the categorization of the events in the HP and LP SRs, to evaluate the $E_{\mathrm{T}}^{\mathrm{miss}}$ and for the MJ estimate.

G. Di Gregorio, *A Beauty-ful Boson*, Springer Theses,
https://doi.org/10.1007/978-3-031-20013-7

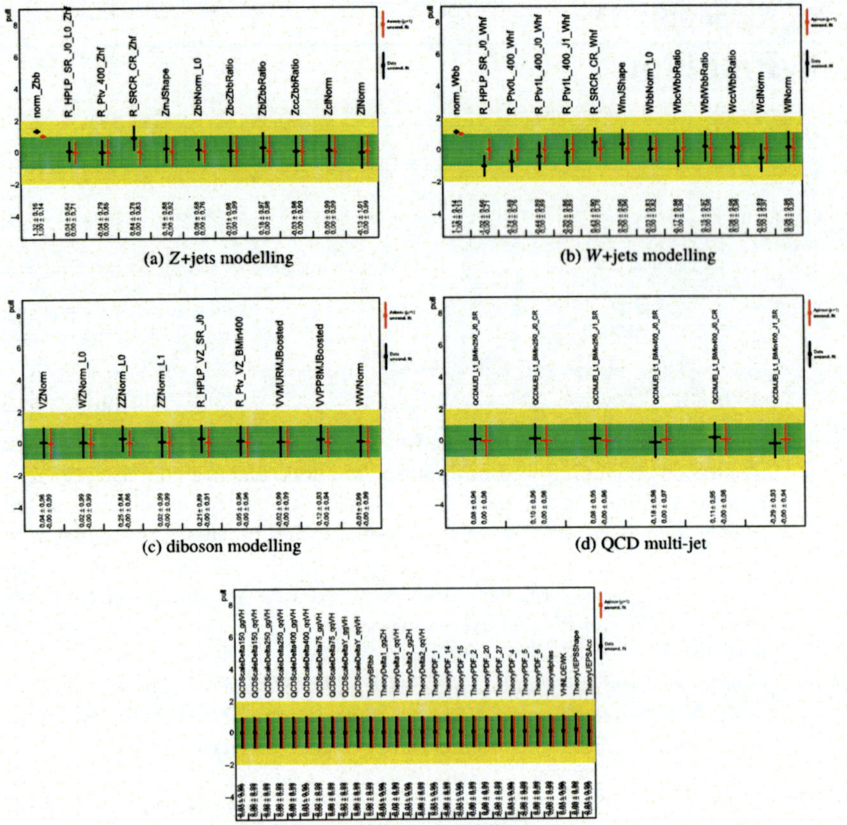

(a) *Z*+jets modelling

(b) *W*+jets modelling

(c) diboson modelling

(d) QCD multi-jet

(e) *VH* signal modelling

Fig. D.1 Monte Carlo modelling-related and QCD multi-jet nuisance parameter pulls, constraints and post-fit normalisation scale factors for the freely floating parameters, obtained from the unconditional fit in all three lepton channels to the Asimov- (*red*) and the actual dataset (*black*)

Figure D.2b shows the NPs associated to the lepton uncertainties. The `EG_RESOLUTION_ALL` and `EG_SCALE_ALL` NPs are related the electron energy resolution and scale uncertainties, while the `EL_` and `MUON_` NPs include the electron and muon uncertainties on the identification, reconstruction and isolation efficiency, respectively. Figure D.2c and d show the NPs related to the $E_{\mathrm{T}}^{\mathrm{miss}}$ and

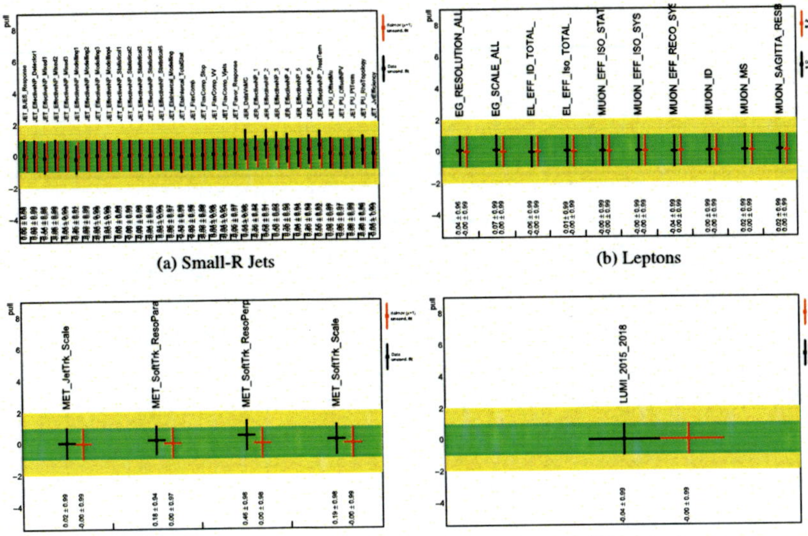

(a) Small-R Jets

(b) Leptons

(c) Missing transverse energy

(d) Luminosity

Fig. D.2 Detector-related nuisance parameter pulls and constraints, obtained from the unconditional fit in all three lepton channels to the Asimov- (*red*) and the actual dataset (*black*)

luminosity uncertainties, respectively. There is no unexpected behaviour for these uncertainties.